Women in Engineering and Science

Series Editor
Jill S. Tietjen
Greenwood Village, CO, USA

The Springer Women in Engineering and Science series highlights women's accomplishments in these critical fields. The foundational volume in the series provides a broad overview of women's multi-faceted contributions to engineering over the last century. Each subsequent volume is dedicated to illuminating women's research and achievements in key, targeted areas of contemporary engineering and science endeavors.The goal for the series is to raise awareness of the pivotal work women are undertaking in areas of keen importance to our global community.

More information about this series at http://www.springer.com/series/15424

Jill S. Tietjen

Scientific Women

Re-visioning Women's Scientific
Achievements and Impacts

 Springer

Jill S. Tietjen
Technically Speaking, Inc.
Greenwood Village, CO, USA

ISSN 2509-6427   ISSN 2509-6435   (electronic)
Women in Engineering and Science
ISBN 978-3-030-51447-1   ISBN 978-3-030-51445-7   (eBook)
https://doi.org/10.1007/978-3-030-51445-7

This Springer imprint is published by the registered company Springer Nature Switzerland AG
The registered company address is: Gewerbestrasse 11, 6330 Cham, Switzerland

To my sister, Laura S. Privalle, PhD, who has demonstrated throughout her life that women can do science.

Foreword

Today's young women have grown up in an era when their mothers have been traveling into space, tending to our medical needs as physicians, or watching over our legal rights in the Supreme Court. But this was not always the case. Throughout most of its history, science in the USA has been the domain of white males. A few hardy women (as noted in subsequent chapters in this book) made their way into science early in the twentieth century, but the Great Depression of the 1930s, World War II and its ensuing G.I. Bill of the 1940s, and the general tenor of the culture in the 1950s all erected barriers to women that relatively few were able to hurdle.

During the second half of the twentieth century, however, many doors to higher education were left ajar and women began to make their way through them in significant numbers to prepare for full-time careers outside the home. They now earn more than half of all bachelor's and master's degrees and are approaching parity at the doctorate level (44% in 2000). And their progress in preparing to join the science and engineering community is remarkable. In the decades of the 1950s, women earned only 15% of the bachelor's degrees, 10% of the master's, and 6% of the doctorates in science and engineering. Some 45 years later in 2000, women had increased their share of degrees earned to 50% at the bachelor's level, 43% at the master's level, and 36% of the doctorates.

Despite this success in preparing themselves for careers in science, women's ability to crack through the glass ceiling in the corporate and academic arenas has been limited. Although the salary gap is narrowing, women continue to earn less

Author's Note: When this manuscript was originally drafted almost 20 years ago (yes, some books take a long time to find their way into print), Eleanor Babco was exceedingly helpful and supportive. She also wrote this wonderful foreword. Even though the information is a little dated, it is still relevant today. I am fortunate to also have a second foreword for this book, written by my friend and fellow inductee into the Colorado Women's Hall of Fame – JoAnn Cram Joselyn.

than their male colleagues. It is only within the last couple of years that a woman reached the top of a technology company when Carly Fiorina became CEO of Hewlett-Packard in 1999. Also, 1999 marked the year that a woman PhD chemist, Cynthia Trudell, became President of Saturn Motors and Shirley Ann Jackson, a PhD theoretical physicist, became President of Rensselaer Polytechnic Institute. And, of the 1,907 active members of the National Academy of Sciences, only 187 are women.

So to the young women of today who aspire to a career as a scientist, read on. The stories of the women who followed their dreams and did science will inspire you to do likewise. Science is exciting, science is nurturing, science is fun, and science is rewarding. As we continue in the new millennium, the doors into the science community are not blocked for women as they were for many in the early part of the twentieth century. We should all work toward making science so inviting for young women that it will not be unusual for them to head large research groups or companies, or win prestigious awards in greater numbers. No nation can afford to abandon any of its talent – particularly the majority half who are women.

All young people should grow up knowing that they, too, can have the fun of solving problems and making discoveries, i.e., do science. As young women read this book, which poignantly tells the story of how their foremothers were truly in love with science, they, too, will start to realize that they can become scientists. For they can.

Eleanor Babco (1942–2013)
Former Executive Director
Commission on Professionals for Science and Technology

Washington, DC, USA   Eleanor Babco

Foreword

In this volume, Jill Tietjen has effectively and succinctly told the story of women who have achieved major scientific breakthroughs, despite overwhelming cultural bias. We see that women can do science (and makes one wonder what might have been achieved had not the barriers been in place). I can add to the story as one who achieved despite recognizable barriers and almost in spite of herself. How? – primarily through the power of the "network" – women and men who provided a push when the time was right, almost like pumping a swing.

My parents were the first to help me out. My Mother, much to her regret, did not get a college education because in the depression-era 1930s, her family could send only one child to college – their son. Fortunately for me, even though I was otherwise raised "pink," mom was determined that I would be properly educated. My civil engineer father recognized and encouraged my natural curiosity and gave me a telescope at age 8. As a teenager, I was especially inspired by Sputnik – that launch, on my 14th birthday, was pivotal. Dad had a short-wave radio and we could hear the beep-beep-beep of the signal when Sputnik passed overhead.

The next push came from my high school math teacher, Mrs. Price. She provided the needed female mentorship to assure me that math and science were OK for girls, and I decided to pursue engineering as a career. I graduated in Applied Math but then went to graduate school in AstroGeophysics because a professor, Dr. Don Billings, had offered me a summer job at Mt. Wilson Observatory doing solar observations. That lucky break led to a Ph.D. (the first woman to earn one in that discipline from the University of Colorado) and a career in space weather science and forecasting. But academic achievement did not provide me with the self-confidence and management skills that eventually made a difference. Those skills first came from non-professional groups – beginning with the Girl Scouts and including women's societies like the Society of Women Engineers, American Association of University Women, the P.E.O. Sisterhood, and many church committees including mission circles. This foundation led to volunteering with professional societies (e.g., the American Geophysical Union, the International Association of Geomagnetism and Aeronomy) that provided the "big push" and led to my election as the first woman (and first American) to become the Secretary General of the

International Union of Geodesy and Geophysics, a non-governmental organization that brings together the scientists of the world in 8 specific disciplines: geodesy (gravity field of the earth), geomagnetism and aeronomy, meteorology, fresh water hydrology, oceanography, seismology, volcanology, and cryospheric science – the study of glaciers and global ice fields.

How did I break the stereotypical bond that stymied (still today) so many young women? My parents were key – they said I could and after that, I could dismiss the negative messages that came my way. Networking then powered the outcome. It's amazing what a little encouragement can do. Push!

JoAnn Cram Joselyn

Secretary General, International Union of Geodesy and Geophysics (1999–2007)

Space Scientist, NOAA Space Weather Predictions Center (1967–1999)

Boulder, CO, USA   JoAnn Cram Joselyn

Preface

In 1987, when Alexis Swoboda and I began researching great women in engineering and science, I only knew one woman who fit into that category – two-time Nobel Laureate Marie Curie. I also had no idea that this was the beginning of my life's journey. Discovering the historical – and current – women in engineering and science around the world, documenting their accomplishments, and nominating them for awards became a labor of love. It was only later that I found out how little had been written about women in science, and about women in every other field of endeavor for that matter, and how invisible and marginalized women were. It became my mission to change that. As Gloria Steinem says, "History and the past are not the same." I am not, as some would say, revising history. I am restoring the narrative – the stories of the half of the world's population who are left out of most of written history.

This volume was originally drafted in the early 2000s for the "Setting the Record Straight" series; a series of books that I wrote in conjunction with Dr. Betty Reynolds. It was never published. When it finally dawned on me in February of 2020 that the Springer Women in Engineering and Science series was about both women in engineering and women in science, and that I had previously published the history of women in engineering through the series, the appropriate venue for this volume's publication became obvious.

During the pandemic of 2020 and the associated "stay-at-home" order in the state of Colorado where I live, the book that had been previously drafted was reconfigured, revised, and updated. It provides a brief overview of how the sciences evolved, from early understandings of astronomy and mathematics through chemistry and physics. And, it documents women's roles in science as that evolution occurred. Along the way, women and their accomplishments are documented. Those accomplishments underlie the quality of life and standard of living that humans around the globe enjoy today.

It has been a joy to relearn about women I already knew about and to discover new ones. It is important to me to be inclusive so that the diversity of women and their accomplishments are incorporated. I have two books published to date in the Her Story series. I believe this book is a worthy addition in that vein as well.

Jill S. Tietjen, PE

Co-author, *Her Story: A Timeline of the Women Who Changed America*

Co-author, *Hollywood: Her Story, An Illustrated History of Women and the Movies*

Greenwood Village, CO, USA   Jill S. Tietjen

Acknowledgments

As with any undertaking of this sort, many people provided support, assistance, and encouragement. A huge thank you to the following and I apologize in advance for any errors or omissions.

To Alexis Swoboda who planted the original seed. I am forever indebted. Her role as editor of this volume is greatly appreciated.

To all of my Society of Women Engineers colleagues (who I consider a family unit), especially Beth Boaz, Yvonne Brill, Sherita Ceasar, Jane Daniels, Patricia Eng, Alma Martinez Fallon, Jude Garzolini, Jamie Ho, Gina Holland, Helen Huckenpahler, Suzanne Jenniches, Tammy Johnson, Connie King, Peggy Layne, Rachel McQuillen, Dorothy Morris, Islin Munisteri, Anne Perusek, Mary Petryszyn, Carolyn Phillips, Ada Pressman, Nancy Prymak, Mary Rogers, Meredith Ross, Anna Salguero, Sandra Scanlon, Kristy Schloss, Nanette Schulz, Jackie Spear, Alexis Swoboda, Mary Ann Tavery, Robin Vidimos, Kitty Wang, Tracey Whaley, and Jere Zimmerman.

A debt of gratitude to Betty Reynolds, Kendall Bohannon, Wendy DuBow, Carol Carter, and Sande Johnson for paving the way.

In particular, Tiffany Gasbarrini and Rebecca Hytowitz at Springer US for their belief, support, and encouragement. I couldn't ask for a better editor than Mary James. Thank you also to Zoe Kennedy and Charles Glaser. Springer's commitment to move forward with the Women in Engineering and Science series is saluted.

To my village (Enid Ablowitz, Gail Berkey, Yolanda Bryant, Lisa Downing, Jennifer Hellier, Glo Martinez, Colleen Miller, Marie Sager), whose love, support, guidance, and, at times, redirection are appreciated more than they know.

To my husband, David Tietjen, who supports all of my many endeavors.

Jill S. Tietjen

Contents

List of Figures

List of Tables

About the Author

Jill S. Tietjen, PE entered the University of Virginia in the Fall of 1972 (the third year that women were admitted as undergraduates after a suit was filed in court by women seeking admission) intending to be a mathematics major. But midway through her first semester, she found engineering and made all of the arrangements necessary to transfer. In 1976, she graduated with a B.S. in Applied Mathematics (minor in Electrical Engineering) (Tau Beta Pi, Virginia Alpha) and went to work in the electric utility industry.

Galvanized by the fact that no one, not even her Ph.D. engineer father, had encouraged her to pursue an engineering education and that only after her graduation did she discover that her degree was not ABET-accredited, she joined the Society of Women Engineers (SWE) and for more than 40 years has worked to encourage young women to pursue science, technology, engineering, and mathematics (STEM) careers. In 1982, she became licensed as a professional engineer in Colorado.

Tietjen starting working jigsaw puzzles at age two and has always loved to solve problems. She derives tremendous satisfaction seeing the result of her work – the electricity product that is so reliable that most Americans just take its provision for granted. Flying at night and seeing the lights below, she knows that she had a hand in this infrastructure miracle. An expert witness, she works to plan new power plants.

Her efforts to nominate women for awards began in SWE and have progressed to her acknowledgment as one of the top nominators of women in the country.

Her nominees have received the National Medal of Technology and the Kate Gleason Medal; they have been inducted into the National Women's Hall of Fame and state Halls including Colorado, Maryland, and Delaware; and have received university and professional society recognition. Tietjen believes that it is imperative to nominate women for awards – for the role modeling and knowledge of women's accomplishments that it provides for the youth of our country.

Tietjen received her MBA from the University of North Carolina at Charlotte. She has been the recipient of many awards including the Distinguished Service Award from SWE (of which she has been named a Fellow and is a National Past President), the Distinguished Alumna Award from the University of Virginia, and the Distinguished Alumna Award from the University of North Carolina at Charlotte. She has been inducted into the Colorado Women's Hall of Fame and the Colorado Authors' Hall of Fame. Tietjen sits on the boards of Georgia Transmission Corporation and Merrick & Company. Her publications include the bestselling and award-winning book *Her Story: A Timeline of the Women Who Changed America* for which she received the Daughters of the American Revolution History Award Medal and the bestselling and award-winning book *Hollywood: Her Story, An Illustrated History of Women and the Movies*.

Chapter 1
Introduction

Women in science in the United States have historically been so few and so unrecognized that the directory listing individuals in the science fields, *American Men of Science*, did not even change its name to the more inclusive, *American Men and Women of Science*, until 1971[1] [1, 2]. But women around the world have been in the various scientific fields for centuries and their accomplishments have been significant. Science generally means a search for knowledge and universal truths for their own sake. Sciences covered in this book include mathematics, physical sciences, computer sciences, life sciences and social sciences. Women have been attracted to the many fields of science and have achieved in these fields in spite of tremendous obstacles placed in their way over the centuries. This book chronicles and celebrates their stories. In the early centures that focus is international. The primary focus in the 1900s and 2000s is women in science in the U.S.

But women around the world have been in the various scientific fields for centuries and their accomplishments are significant. By 1759, botanist Jane Colden had become the first person to identify and describe the gardenia. Ada Byron Lovelace foresaw the development of computer software in 1843. Astronomer Maria Mitchell discovered a comet in 1847. In 1925, Florence Sabin became the first woman elected to the National Academy of Sciences. That Academy had been established in 1863 [3–5].

These accomplishments and those of many other women have come in spite of the sentiments like that expressed below.

Culture in young women should never develop into learning; for then it ceases to be delicate feminine culture. A young woman cannot and ought not to plunge with the obstinate and preserving strength of a man into scientific pursuits so as to become forgetful of everything

---

[1] The R.R. Bowker Company of New York announced in November 1971 that *American Men of Science,* which had been published periodically since 1906, would be renamed *American Men and Women of Science*. That first 1906 edition included 149 women out of a total of 4131 entries.

J. S. Tietjen, *Scientific Women*, Women in Engineering and Science,
https://doi.org/10.1007/978-3-030-51445-7_1

else… Only an entirely unwomanly young woman could try to become so thoroughly learned, in a man's sense of the term; and she would try in vain, for she has not the mental faculties of a man…. Author unknown [6].

1.1   What Is Science?

Science is derived from the Latin *scientia*, knowledge, the present participle of *scire*, to know. Science generally means a search for knowledge and universal truths for their own sake – not for the sake of the application of such knowledge, which would generally be considered technology or engineering [7]. An alternative definition for science, that explains how this search for knowledge and universal truth is pursued, is an attempt to explain natural phenomena through seeking agreement between observational data and theoretical assumptions [8].

By today's definitions, the first real science would have been what Aristotle (384–322 B.C.) designated as *physis* ("the nature of things"). The basic scientific principles as we know them in the modern western world began to be developed during the period 1300 to 1700 [7]. Today, what is called science is actually many fields of endeavor including architecture, engineering, surveying, mathematics, physical sciences, computer sciences, life sciences, social sciences, and medicine and health. The sciences covered in this book will include all of these except architecture, engineering, surveying and medicine and health. Occupational titles and categories within each of the covered categories are included in Appendix A [9].

1.2   Women Scientists

Women have contributed to the myriad of fields of scientific endeavor throughout time, although many of the records are obscured or lost. Women's scientific participation has occurred in spite of enormous obstacles characterized as "women have been more systematically excluded from doing serious science than from performing any other society activity except, perhaps, frontline warfare" [10].

French mathematician, Sophie Germain, was determined to study mathematics. This determination persisted even though her parents made sure her bedroom was without light or fire and that she was left without clothes. It persisted in spite of the fact that she was not allowed into the École Polytechnique. Her foundational work on Fermat's Last Theorem was published by her mentor and is still referred to in textbooks as Germain's Theorem. Her work on vibrations and elasticity won her a prize from the French Academy of Sciences (1816), but her paper was not published until after her death [3, 4].

Nobel Laureate Maria Goeppert-Mayer (Nobel Prize in Physics, 1963) at age 53, ten years after the discovery that would win her the Nobel Prize, was finally offered a regular, full-time, paid university job – as a full professor. Up to that point, she had worked in converted closets, lectured and performed research without pay, or served as part-time faculty with gratuitous pay. Yet, she loved physics and was awarded the Nobel Prize in Physics for developing the nuclear shell theory of the atom explained by spin-orbit coupling [10, 11].

Jewel Plummer Cobb had to fight two forms of prejudice – sexism and racism – to achieve in her field of biology. When she was at the University of Michigan, women weren't allowed to walk into the front door of the men's union building and African-American students lived in segregated housing and were not allowed into various restaurants. A single mother after her divorce, Cobb became the president of California State University at Fullerton in 1981 [7, 12].

Women have been attracted to the many fields of science and have achieved in these fields in spite of tremendous obstacles placed in their way over the centuries. This book chronicles and celebrates their stories and accomplishments. In the early centuries, the book's focus is international. The primary focus in the 1900s and 2000s is women in science in the U.S.

References

1. M.W. Rossiter, *Women Scientists in America: Before Affirmative Action, 1940–1972* (The Johns Hopkins University Press, Baltimore, 1995)
2. American Men and Women of Science. https://en.wikipedia.org/wiki/American_Men_and_Women_of_Science. Accessed 7 Apr 2020
3. P. Proffitt (ed.), *Notable Women Scientists* (The Gale Group, Detroit, 1999)
4. C. Morrow, T. Perl (eds.), *Notable Women in Mathematics: A Biographical Dictionary* (Greenwood Press, Westport, 1998)
5. National Academy of Sciences: Mission. www.nasonline.org/about-nas/mission/. Accessed 7 Apr 2020
6. M.W. Rossiter, *Women Scientists in America: Struggles and Strategies to 1940* (The Johns Hopkins University Press, Baltimore, 1992)
7. S.A. Ambrose, K.L. Dunkle, B.B. Lazarus, I. Nair, D.A. Harkus, *Journeys of Women in Science and Engineering: No Universal Constants* (Temple University Press, Philadelphia, 1997)
8. M.B. Ogilvie, *Women in Science: Antiquity through the Nineteenth Century, a Biographical Dictionary with Annotated Bibliography* (MIT Press, Cambridge, 1993)
9. U.S. Department of Labor, *Dictionary of Occupational Titles*, Fourth Edison, Revised 1991, vol. 1
10. S.B. McGrayne, *Nobel Prize Women in Science: Their Lives, Struggles, and Momentous Discoveries* (Carol Publishing Group, New York, 1993)
11. B. F. Shearer, B. S. Shearer (eds.), *Notable Women in the Physical Sciences* (Greenwood Press, Westport, 1997)
12. B. F. Shearer, B. S. Shearer (eds.), *Notable Women in the Life Sciences* (Greenwood Press, Westport, 1996)

Chapter 2
History of Science

2.1 The Beginnings of Science

Western science has its roots in Greece. However, two of the most important tools used to express science – writing and mathematics – were developed in Mesopotamia and Egypt[1] [1, 2]. Early forms of writing emerged between 3500 and 3000 BC. The two distinct forms were the cuneiform script of Mesopotamia which were on clay tablets and the hieroglyphics of Egypt inscribed on papyrus, both of which were syllabic, not alphabetic[2] [2–4]. By about 2400 B.C., the Sumerians in Mesopotamia had developed a numbering system in cuneiform. This system had evolved from bits of stuff used as tokens, i.e., the sale of 62 farm animals formalized by placing 62 tokens inside a clay container, into a system where the position of the number had significance, called positional notation. Interestingly, the Mesopotamian numbering system, and later versions of it within the region, used a base of 60, as compared to our modern base of 10. Vestiges of the Mesopotamian system remain with us today, however, in the form of 60 min to the hour, 60 s to the minute, and 360° to the circle [5].

Astronomy is considered the oldest of the natural sciences originating about 5000 years ago. It became the first science as the writing and mathematics developed by the Mesopotamian and Egyptian civilizations were applied in a very practical manner – to develop a calendar, noting the progression of the seasons. The Egyptians were responsible for the 365-day year, twelve months of 30 days each

---

[1] Mesopotamia, the land "Between the Rivers" occupies the flat alluvial area between the Tigris and the Euphrates Rivers in what is now Iraq. The "Fertile Crescent" includes this area plus the area north and west which stretches in a curve from the Tigris River around to the coasts of Syria, Lebanon, and northern Israel.

[2] For a syllabic writing system, each sign represents a syllable. Alphabets are comprised of letters, each one of which represents a single sound. Letters are then combined to form syllables.

© The Editor(s) (if applicable) and The Author(s), under exclusive license to
Springer Nature Switzerland AG 2020
J. S. Tietjen, *Scientific Women*, Women in Engineering and Science,
https://doi.org/10.1007/978-3-030-51445-7_2

(originally based on the lunar cycle with extra days added as necessary), and a 24-hour day [2, 3, 5].

Mathematics and language continued to develop as well. By 1800 B.C., the Mesopotamians had developed geometry and could solve binomial equations [5]. Mathematical tables were developed. These included multiplication, division, squared numbers, and square roots tables [4]. Around 1800 to 1500 B.C. the first alphabetic writing systems emerged, labeled Ugaritic and Phoenician [2].

These early civilizations used science, including mathematics, to facilitate record-keeping and to organize their lives. The Greek civilization from 600 B.C. to 500 A.D., however, opened new frontiers in science by asking not just how the world worked, but why [5]. The Greeks looked for general patterns in nature, to find fundamental, orderly, and systematic principles [4, 5]. Because they were primarily philosophers, however, and not scientists, much of their effort consisted of subjective thought and intellectual exercise – very little consisted of observation or experiment [4].

One of the key products of the Greek civilization was the idea that the Earth was at the center of the universe; an idea that would endure for centuries. The introduction of free inquiry and vigorous debate is also a legacy of the Greek civilization [5].

Numerous advances in science are attributable to this period. Archimedes (c. 287–212 B.C.) calculated the value of pi (π) and made many other significant contributions to geometry as did Euclid (330–260 B.C.)[3] [3–5]. Alexander the Great founded the city of Alexandria in Egypt with its famous Library of Alexandria [4]. The Pythagorean Theorem, named for Pythagoras (560–480 B.C.), defined relationships in right triangles [4, 5]. Pythagoras was the first to determine that the Earth was a sphere and his view was later reinforced by Aristotle. (384–322 B.C.) [5].

The medieval period from about 500–1000 A.D. is often referred to as the Dark Ages. However, scientific advances did occur during this period and scientific knowledge that had existed in the Greco-Roman cultures was preserved and enhanced by the Islamic Byzantine Empire [5]. One particularly significant advance that occurred during this time was the adaptation of an Indian numbering system that incorporated true positional notation and a symbol for zero, now called Arabic numbers[4] [2–5]. Scientific inventions from China, including paper, available about

---

[3]Euclid's *Elements* became the most enduring and widely-studied secular book in the western world, dominating the teaching of mathematics for more than two thousand years. Principally devoted to plane and solid geometry, Euclid's work provides numerous theorems and problems in the classic form of definition, demonstration, and proof, which came to be seen as a paradigm of logical thought. Euclid provided a language and method of mathematics capable of application to astronomy, optics, mechanics, and engineering.

[4]The Hindus in India originally developed a numbering system based on a decimal system as far back as the Vedic period (1500 B.C.). Arabic numerals were promulgated by the famous ninth-century Arab mathematician Al Khwarizmi (780–850) from whose name the word "algorithm" is derived. Al Khwarizmi also used algebra and it is from one of his works that the word itself is derived – "The Book of *al-Jabr* and *al-Muqabala*" meaning restoration and balancing. Arabic numerals entered Europe 300 years later when Adelard of Bath began translating Arabic works into Latin. Fibonacci (c. 1170–1240 also known as Leonardo of Pisa) taught and wrote using Arabic

1150 A.D. in Europe, and the compass were adopted by western civilization[5] [2, 4, 5]. Alhazen (965–1040), a physicist from what is now Iraq, investigated vision and became the father of modern optics [5]. Persian astronomer al-Sufi (903–986) fixed the names of several hundred stars, many of them still in use [3].

Some renaissance of learning was begun during Charlemagne's reign in the eighth and ninth centuries [3]. The fall of Toledo, Spain in 1085, led to the recovery of Greek and Islamic science by western scholars who began to translate the works from Arabic into Latin[6] [1, 3]. The founding of a university in Paris around 1170 followed by the founding of a university at Oxford led to more formal dissemination of Western European knowledge. The Franciscans, who were teaching at Oxford by the 1220s, set about learning mathematics and natural science with knowledge gleaned from the translated documents[7] [2].

The Renaissance, the embracing of humanistic beliefs and values, began in Italy in the fourteenth century. It was during this time that a general change came about in the way in which man viewed himself and the world in which he lived [2]. By 1450, when Johannes Gutenberg invented movable type, enabling large scale printing as opposed to hand-written manuscripts (usually the products of monasteries), the stage was set for the scientific revolution[8] [2, 5].

The Scientific Revolution of the sixteenth and seventeenth centuries affected every field of science, changed the techniques of scientific investigation, changed the goals that a scientist set for himself, and changed the role that science played in philosophy and in society itself [2]. Mathematics became a professional tool with the first printed text of Euclid being published in 1482 [3]. The definitive statement of sixteenth-century knowledge of geology, metallurgy, and mine-engineering, *De re metallica*, was published in 1556. From 1530 onward, a series of works began to be published that more objectively and completely classified plants and animals [3].

---

numerals and his work *Liber Abbaci* was the first western mathematical treatise to make systematic use of the numerals explaining the concept of place value where a digit could represent units, tens, hundreds, or thousands. Fibonacci is more commonly known today for the Fibonacci sequence of 1, 2, 3, 5, 8, 13, 21, 34, and so forth where each number is the sum of the two preceding numbers, a sequence he was the first to describe.

[5] The Chinese also invented gunpowder and developed silk, all of which were imported to the West in the seventeenth century after navigational advances broke through Chinese isolation. The Chinese also had superior astronomical observations including the supernovas of 1006, 1054, 1572, and 1604. They recorded eclipses back to 720 B.C.,75 "guest stars" between 352 B.C. and 1604 A.D., and comets over 22 centuries. They were the first to systematically catalog the stars in the skies. Their scientific mapmaking was significantly advanced over Western mapmaking and by the twelfth century A.D., the Chinese had already realized that mountains were elevated land masses that had once formed the sea floor.

[6] Alfonso VI of Castille captured Toledo from the Arabs.

[7] A Catholic order founded in 1209 by Saint Francis of Assisi, the Franciscans are often referred to as Greyfriars.

[8] The invention of printing illustrations from engraved metal plates which originated in the Rhine valley and in northern Italy in the 1450s also helped spread the knowledge in Renaissance times and was to have special importance to some scientific fields.

But it was the Copernican revolution (the Earth and other planets revolved around the Sun) that fundamentally altered how man viewed himself and his place in the world[9] [3–5]. Since the Earth was no longer the center of the universe, two broad tendencies began to emerge: (1) the ancient authorities, including a particularly revered Aristotle, were evidently wrong about many of their assumptions and (2) God began to retreat from the day-to-day events of the world as it appeared that the world operated on scientific and invariable principles [5]. Additional work and observations by Tycho Brahe (1546–1601) who was assisted by his sister Sophia Brahe (1556–1643), Johannes Kepler (1571–1630), and Galileo Galilei (1564–1642) in the late 1500s and early 1600s, and significantly assisted by the invention of the telescope in the early 1600s, would not only advance the science of astronomy and celestial mechanics but would further dispute classical assumptions that had been in place since the time of Aristotle[10] [3].

Galileo advocated the quantitative study of nature, advanced the design of the microscope, and advanced some theories of motion and mechanics. Mathematical analysis was the hallmark of the new science [3–5]. Biology made tremendous strides about this time as well. The understanding of the value and usages of the microscope, described around the 1630s, led to the discovery of protozoa, bacteria, and many other tiny creatures. A study of insects, facilitated by the use of the micro-scope, was published by Jan Swammerdam (1637–1680). Plants were clearly described in a 1682 publication in which the microscope had been applied [3]. To help spread information about scientific developments, the Royal Society of London for the Promotion of Natural Knowledge (1662) and the Academy of Sciences in Paris (1666) were established. The Academy of Sciences in Berlin was established in 1700 [2, 5].

Mathematics advanced with concepts of negative numbers, decimal fractions, the development of logarithmic tables, and improvements in algebraic languages. The stage was now set for Sir Isaac Newton (1642–1727) who revolutionized the scientific field of physics [3]. Newton developed calculus to provide a mathematical framework to deal with constantly changing values, set forth the theory of gravity, determined the laws of motion published in *Philosophiae naturalis principia mathmetica* (1687 – known as *Principia*), and developed the reflecting telescope [5]. His *Principia* was translated into French by Emilie Du Chatelet (1706–1749) [3, 4].

---

[9]Copernicus (1473–1543) pursued astronomical studies in addition to his studies in law and medi-cine. He developed his theory of the solar system being heliocentric (with the sun in the middle) to explain inconsistencies in the Ptolemy scheme of the Earth being the center of the universe that just didn't work. Copernicus did not deviate from Ptolemy's theory that the planets orbit in circles (this would be fixed by Johannes Kepler), but he did calculate that Mercury and Venus were closer to the Sun than the Earth. His work was completed by 1510, yet he did not publish it officially until 1543, after one of his colleagues announced the Copernican system. His theories were not accepted for many years.

[10]The development of eyeglasses in northern Italy in the late 1200s, followed by improvements in lenses led to the first telescopes in the Netherlands in 1608. Galileo invented an improved telescope in 1609 and published initial astronomical observations made with his telescope in 1610 [6–8].

The birth of modern chemistry was prodded along by Sir Robert Boyle (1627–1691) and Robert Hooke (1635–1702). Boyle's book *The Sceptical Chymist* proposed a system of elements that would later be refined into the periodic table[11] [5]. The first reliable thermometer was produced in 1709 by German Daniel Gabriel Fahrenheit (1686–1736) [5]. Additional significant advances in chemistry would come in the mid to late 1700s including the naming of oxygen and hydrogen by Antoine-Laurent Lavoisier (1743–1794), assisted by his wife Marie (1758–1836). Lavoisier is now regarded as the father of modern chemistry [2, 5].

Geology moved from the realm of opinion and speculation to science during the seventeenth and eighteenth centuries. Paleontology – the study of fossil plants and rock – began to receive careful study. Rene Descartes (1596–1650) advanced suggestions about the composition of the earth in 1617.[12] The principles of modern physical geology were put forth in Niels Stensen's *The Prodromus to a Dissertation Concerning Solids Naturally Contained Within Solids* in 1671. Abraham Werner founded a geological school in Germany in 1775 and made important contributions with regard to minerals [2].

Classification systems for plants, animals, and minerals were put forth by Swedish naturalist Carl von Linné (1707–1778), now better known as Linnaeus. This system included the two-name system where one name gives the genus or common characteristic and the other the species (Fig. 2.1) [2].

Physics advanced significantly in 1752 when Benjamin Franklin (1706–1790) discovered that lightning was a form of electricity (Fig. 2.2) [5]. The first steam engine was developed in 1698 and then improved upon significantly by James Watt (1736–1819) in 1765 (Fig. 2.3) [3, 5]. Economics made a major leap forward with the publication of Adam Smith's (1723–1790) *Wealth of Nations* in 1776 [3].

The pace of scientific discoveries and innovations would only increase throughout the 1800s. But, with few exceptions, the work of women scientists and knowledge of their contributions from antiquity through the 1800s remained a significant rarity, usually existing only on the periphery or intertwined with the contributions of a family member, usually a husband or brother. Rarely did a woman continue any scientific interests once she married. The persistence of women's subordinate status and the idea that women were fundamentally different than men was not conducive to women contributing to scientific, political, or social advances [1].

---

[11] Boyle's Law (1662) defined the relationship between pressure and volume. Hooke's law provides the rule of deformation.

[12] Descartes suggested in *Principles of Philosophy* that the Earth was once molten, like the Sun, but had now cooled and condensed except in its central regions.

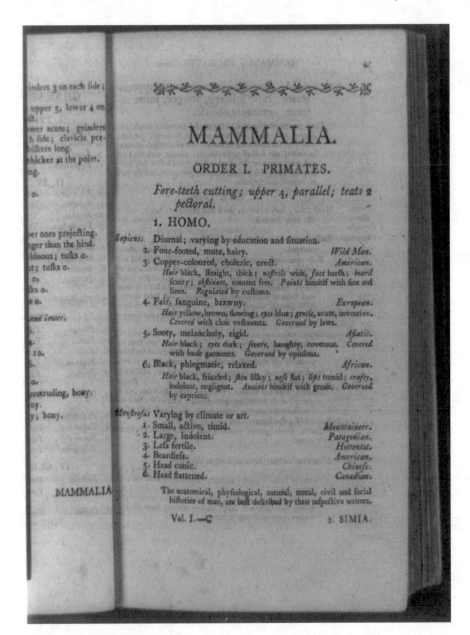

Fig. 2.1 Linne's Mammalia. (Courtesy of Library of Congress)

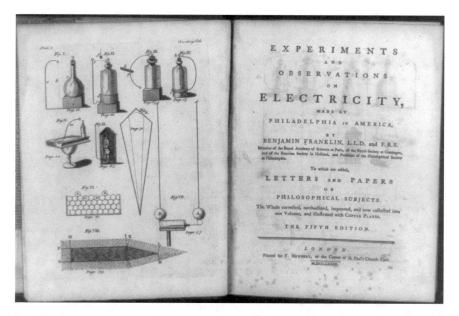

Fig. 2.2 Benjamin Franklin's documentation of his electricity experiments. (Courtesy of Library of Congress)

2.2   The 1700s

Education was not a widespread privilege (or requirement) for many years after our nation's founding. Formal education was rare, public schools as we know them today did not exist, and what education was available was not free. Some families hired tutors for their sons; rarely could their daughters even join in. Lecturers traveled around the country orating on topics of interest.

In fact – to be formally educated – especially beyond the high school level – was a rare privilege throughout most of Western history and exclusively granted to affluent males. When the Constitution was adopted, few colonials had attended school – the literacy rate for white women was about 40% and for white men about 80% [9, 10].

At this time in U.S. history, pursuing any form of rigorous education was considered inappropriate for women; they might harm their reproductive capabilities especially if they filled their heads with radical ideas [10]. In addition, because the laws and customs that took hold first in the original thirteen colonies were based on English Common Law, women's social status was acquired by birth or marriage. Women were deemed subordinate to men and were expected to play a subservient role, first to their fathers and then to their husbands [11]. Women had many duties befitting their station, but few, if any rights; they were trapped in a condition known now as "civil death" [9, 11].

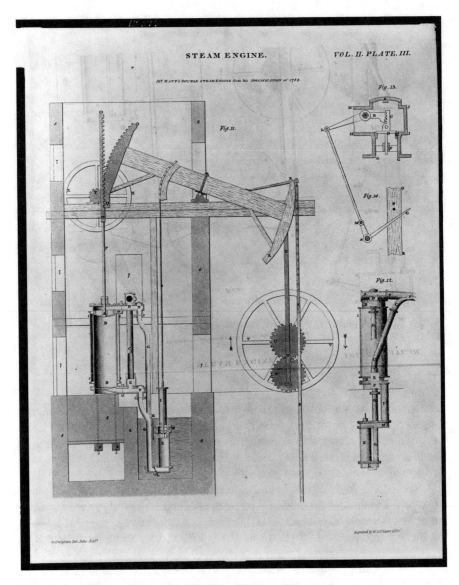

Fig. 2.3 James Watt's steam engine. (Courtesy of Library of Congress)

Women's sphere was narrowly defined as domestic, and the female role specifically defined as wife and mother. Conditions were even worse for minority women. While they were no doubt involved in some form of "science" and "technology" long before, during, and after the Colonial period, there is scant documentation to confirm their contributions. With the exception of limited reports of black women who engaged in medical practices on the plantations, there is little else to suggest that scientific pursuits were within the purview of minority women until long after the Civil War [12].

By the end of the eighteenth century, however, ideas about female education reform were beginning to draw attention. Essayist, poet and playwright Judith Sargent Murray protested the lack of equality in education between boys and girls in her 1790 tract *On the Equality of the Sexes* [9, 11, 13]. Female private schools, female seminaries or "dame schools," had begun to spring up in the 1780s, although attendance tended to be limited to the well-to-do. Most of these schools focused on domestic subjects to ensure that their graduates would attract proper suitors [14]. Rarely did schools for boys teach science at this point in history. Thus, it is not surprising that those schools that existed for women primarily included courses on languages, history, geography, and morality [1].

But women were starting to become frustrated with their prescribed position in life. Women read treatises on physics and chemistry. They attended lectures on physics, chemistry, and natural history. Dictionaries and encyclopedias intended for the use of both sexes, publications especially for women, and public lectures, especially those for women, increased significantly in number during the late 1700s and early 1800s – spreading from Philadelphia to Albany, New York City, and Wilmington, Delaware. In 1817, fifty women in Charleston got together to study botany. Women in Georgia City raised $500 to buy themselves a telescope and formed the Caroline Herschel Society [1, 12, 15]. Change was coming.

2.3   The 1800s Prior to the Civil War

Women needed to lead the charge to ensure that education was available to women. Two of the more prominent of these women were Emma Willard and Mary Lyon. Credited with being the first person to make secondary education available for women, Emma Willard was able to inspire the citizens of Troy, New York to raise enough money to build the Troy Female Seminary in 1821. More than anyone else, Willard wrought the basic revolution in the nation's attitude toward the education of women between 1819 and the 1830s [15, 16]. Furthermore, Willard supported teaching natural philosophy to girls, even though some editing was required.[13] Her school, and others like it, essentially became the starting point for women in science and the professions. Willard said [1]:

> Why should we be kept in ignorance of the great machinery of nature, and left to the vulgar notion, that nothing is curious but what deviates from her common course? If mothers were acquainted with this science, they would communicate very many of its principles to their children early in youth... A knowledge of natural philosophy is calculated to heighten the moral taste, by bringing to view the majesty and beauty of order and design; and to enliven piety, by enabling the mind more clearly to perceive, throughout the manifold works of God, that wisdom, in which he hath made them all.

---

[13] Natural philosophy was the study of nature and the universe before the development of what is considered modern science [17].

In some of the sciences proper for our sex, the books written for the other would need alteration; because in some they presuppose more knowledge than female pupils would possess; in others, they have parts not particularly interesting to our sex, and omit subjects immediately relating to their pursuits.

Willard's sister, Almira Hart Lincoln (later Phelps) was also of the opinion that women needed to learn science. Lincoln and Willard both attended lectures of science popularizer Amos Eaton at Rensselaer School (later the Rensselaer Polytechnic Institute).[14] Lincoln began to write textbooks in the 1820s. Her first and most successful, *Familiar Lectures on Botany* (1829), went through at least seventeen editions. She wrote books on other topics including chemistry and natural philosophy helping both to kindle interest in science among men and women in the U.S. and improving the science teaching that was available [15].

Education reformer Mary Lyon (Fig. 2.4) established Mount Holyoke Seminary (later College) in 1837. She had successfully endowed a school for women; and the College exists today as a testament to her efforts. Lyon taught chemistry at Mount Holyoke and stressed science in the curriculum[15] [15, 16].

Also in 1837, Oberlin College set a milestone in education by becoming the first institution of higher education to admit women and students of all races. Oberlin had been established in 1833 as a seminary for men, but later became a college. Women were viewed as a "civilizing influence" on the men and, at first, were actually not allowed to take the same courseload as their male colleagues due to their "smaller brains." By 1841, however, women were allowed to obtain the same bachelor's degrees with the same coursework as the men [10, 11].

Informal education in science for women in the early to mid 1800s was aided by popular books and textbooks, often on botany and written in England, designed for female readers. The most widely read of the foreign scientific books for women were those of the British popularizer Jane Marcet whose topics included chemistry and botany [15].

During the 1840s through the 1860s, the so-called "Age of Reform," women fought for change in many areas including education. Abolition, the right to vote, equal rights for women, as well as improving educational opportunities for women were all areas that women activists were pursuing during this period [11]. And their efforts were starting to yield results.

By 1850, most cities had public schools – at least one for girls and several for boys. The state of education for minorities did not yet measure up to even these standards. And it took the better part of the nineteenth century to expand the free education system for males from elementary schools though high school. By 1860, there were only about 40 schools that qualified as high schools in the entire country [10, 11].

---

[14] Eaton was a self-taught geologist, chemist, and botanist who was particularly adept at the teaching of science to popular audiences through lectures and textbooks. He encouraged Lincoln to publish her textbooks.

[15] Interestingly, Lyon was also influenced by Eaton. She spent a summer at his house in Troy being taught the elements of chemical experimentation.

Fig. 2.4   Mary Lyon. (Courtesy Library of Congress)

Colleges had been established as early as the 1600s in the colonial states and by
the 1800s were primarily found in the eastern U.S. Young men had an opportunity
to attend Harvard College and other Eastern all-male institutions. Although some of
the early female seminaries called themselves "women's colleges," they did not
measure up to these Eastern all-male institutions. However, they did lay the foundation for the establishment of Antioch in 1852, Vassar in 1861, Smith and Wellesley
in 1875, Bryn Mawr and Baltimore College for Women (later Goucher College) in
1885, and Barnard College in 1889, established by Columbia University as its coordinate college for women[16] [10, 11, 15, 16, 18]. And they provided employment for
women. In 1873, of the 2124 persons employed on the faculties of 223 women's

---

[16] Smith College was chartered in 1871 and opened in 1875 with 14 students. Vassar College was
founded in 1861 and opened its doors to its first class of 353 students on September 26, 1865. A
coordinate college means that a separate women's college is established. Although there are links
between the men's and the women's college, the education itself is segregated by gender.

institutions, over 73% were female and an estimated 400 of these women are believed to have been teaching science [15].

The picture was a little different in the Western U.S. The Cherokee National Female Seminary was founded in 1851. Most western colleges were state-supported and usually coeducational from the time of their founding because males were not enrolled in sufficient numbers to support them otherwise and taxpayers would not support the institutions unless their daughters could enroll. Many of these institutions came about as a result of the Morrill Act of 1862 that has been credited with democratizing higher education and providing colleges for the industrial classes[17] [12, 19]. Negro land-grant colleges for the 17 southern states were established by the second Morrill Act of 1890. Mississippi had already established Alcorn University in 1871 under the 1862 Morrill Act. Virginia and South Carolina had shared their funds since 1872 with black colleges in their states [20].

By 1870, Wisconsin, Michigan, Missouri, Iowa, Kansas, Indiana, Minnesota and California had established coeducational state universities. Science curricula at the land-grant institutions were developed and emphasized based on their pertinence to scientific agriculture. The rankings with importance to agriculture were established as [10, 20, 21]:

1. Chemistry
2. Geology and meteorology
3. Botany and horticulture, zoology and biology, and entomology.

Science continued to change at a significant pace during the 1800s. The word "scientist" itself had been coined in 1840 in Glasgow. The American Association for the Advancement of Science (AAAS) was established in 1848. The AAAS, following in the footsteps of the British Association for the Advancement of Science, was founded because of the growing realization of the significance of scientific works and the need to communicate this science to the public. When Charles Darwin's *Origin of Species* was published in November 1859, the entire printing sold out on the first day of publication (Fig. 2.5) [2]. The developments in thermodynamics, the theory of matter, electricity and magnetism, as well as the theory of evolution, advances in geology, and the foundation of the science of genetics significantly advanced human understanding of the world in which we live [1].

Thus, science was evolving and the educational system was evolving as well. The number of women going to college increased dramatically between 1860 and 1920 as educational opportunities became available and women saw the economic and personal benefits of becoming educated [10].

---

[17]The Morrill Act of 1862 signed into law by President Lincoln gave 10,000 acres of Federal government land to each state to sell and use the proceeds to create a public university to teach agriculture and the mechanic (engineering) arts. The land grant universities today still have the major responsibility for agricultural research and teaching responsibility as well as major "outreach" or extension education mission to the public.

Fig. 2.5 Charles Darwin's
The Origin of the Species.
(Courtesy of Library of
Congress)

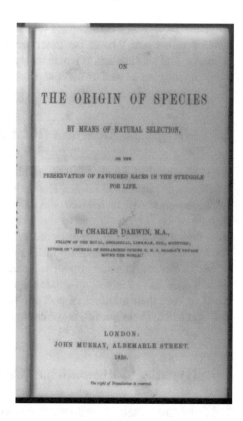

***

KEY HISTORICAL WOMEN

***

The scientific and engineering women from antiquity through the Civil War for whom records survive are quite small in number. The contributions of these women are fascinating and in many cases, enduring. The following brief biographies provide a flavor of the lives they led and the accomplishments credited to them as well as the difficulties they encountered.

2.4   Tapputi (Also Tapputi-Belatekallim): Perfumist (Second Century BC)

Considered the world's first chemist, Tapputi, made perfume and is mentioned on a cuneiform tablet from the second millennium BC in Babylonia. Her perfume contained flowers, oil, calamus, cyperus, myrrh and balsam to which she added water.

This mixture was then distilled and filtered in her still; the oldest referenced still of which we are aware. Women perfumers used the chemical techniques of distillation, extraction and sublimation to create their perfumes, which were important in medicines and religion as well as for cosmetics [22].

2.5   Miriam the Alchemist (First or Second Century AD)

Born in Alexandria, Egypt, Miriam was also known as Mary, Maria, and Miriam the Prophetess or Miriam the Jewess. Her major inventions and improvements included the three-armed still or *tribikos*, the *kerotakis*, and the water bath. Although the purpose of the inventions was to accelerate the process of metals transmuting into gold, they are used extensively in modern science and contemporary households. The *tribikos* was an apparatus for distillation, a process of heating and cooling that imitated processes in nature. Sponges formed a part of the mechanism and served as coolers. The *kerotakis* was an apparatus named for the triangular palette used by artists to keep their mixtures of wax and pigment hot. The water bath, also known as Marie's bath (*bain-marie*), is similar to the present-day double boiler [23].

2.6   Hypatia: Mathematician (Circa 360–415)

The daughter of Theon, a well-known mathematician in Alexandria, Egypt, Hypatia was raised by her father to be a "perfect human being" in spite of the fact that she was a daughter and not a son. Raised to seek knowledge, she was educated in the arts, sciences, literature, philosophy, and all manner of sports. After her mathematical knowledge surpassed that of her father, she was sent to Athens to study. When she returned to Alexandria, she became a teacher of mathematics and philosophy. Hypatia wrote a number of treatises in algebra including significant information on cones being divided by planes. Her inventions included a plane astrolabe, a device used for measuring the positions of the stars, planets and the sun, and to calculate time and the ascendant sign of the zodiac; an apparatus for distilling water, a process used for distilling sea water that is still used today; a graduated brass hydrometer for determining the specific gravity (density) of a liquid; and a hydroscope, a device used to observe objects that lie far below the surface of the water. Her brutal murder led to the end of the formal study of mathematics in Alexandria for over 1000 years [22, 24].

2.7   Hildegard of Bingen: Natural Philosopher (1098–1179)

A German Benedictine abbess known as "the Sibyl of the Rhine," Hildegard of Bingen (Fig. 2.6) wrote music as well as treatises on science including cosmology, medicine, botany, zoology, and geology. Two of her manuscripts, *Causae et curae*

Fig. 2.6 Hildegard of Bingen – Wikipedia

(*Causes and Cures* or *Book of Compound Medicine*) and *Physica* (*Natural History* or *Book of Simple Medicine*) are considered among the greatest scientific works of the Middle Ages and have survived intact. *Physica* is her natural history textbook and included descriptions of nearly 500 plants, metals, stones and animals, and explains their medicinal value to humans. The book became a medical school text. In *Causae et curae*, Hildegard describes the relationships between the macrocosm and specific diseases of the microcosm, the human body, and prescribed medicinal remedies. Hildegard was the first medical writer to stress the importance of boiling drinking water [12, 25, 26].

2.8 Sophia Brahe: Astronomer (1556–1643)

Danishwoman Sophia Brahe, one of ten Brahe children, was educated in classical literature and was a self-taught astrologer and alchemist. She shared those interests with her brother, Tycho, who was ten years her senior. She frequently assisted Tycho with his astronomical observations including those that led to his computation of the lunar eclipse of December 8, 1573. In adulthood, Sophia studied genealogy, astrology, chemistry, botany, and medicine. Although her specific contributions to her brother's astronomical observations are not known, their observations enabled Kepler to determine the elliptical orbits of the planets and their accurate measurements of the positions of the sun, moon, stars, and planets and helped lay the foundation for modern astronomy [1, 25].

2.9   Marie Meurdrac: Alchemist (c. 1610–1680)

Marie Meurdrac was not aware of Miriam the Alchemist's chemistry work when she wrote a six-part chemistry treatise. Meurdrac covered laboratory principles, apparatus and techniques, animals, metals, the properties and preparation of medicinal simple and compound medicines, and cosmetics. Her work included a table of weights as well as 106 alchemical symbols. Her work titled *La Chymie charitable et facile en faveur des dames* was first published in Paris in 1666 (Fig. 2.7). Later editions were issued in 1680 and 1711. Her foreword to her book contained the following thought: … *that minds have no sex and that if the minds of women were*

Fig. 2.7 Marie Meurdrac
Chemistry
Book – Wikipedia

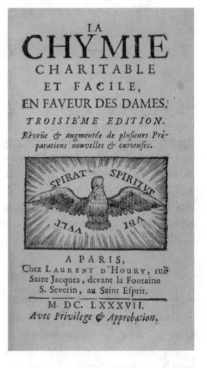

cultivated like those of men, and if as much time and energy were used to instruct the minds of the former, they would equal those of the latter [22].

2.10 Margaret Cavendish, Duchess of Newcastle: Science Popularizer (1623–1673)

Margaret Cavendish was one of the first female scientific writers as a popularizer of science. She was an eccentric and flamboyant woman and one of the most visible of the science ladies of the seventeenth century. She wrote about 14 scientific or quasi-scientific books on natural philosophies and atomism.[18] She was the first woman invited to visit the prestigious Royal Society of London (no other woman was admitted until 1945).[19] Although her significance as a scientist may be minimal, she was a visible advocate of educating women and the author of the first major autobiography written by a woman [1, 12, 25].

2.11 Émilie de Breteuil du Châtelet: Physicist, Chemist, Translator (1706–1749)

The Marquise du Châtelet was tutored as a young woman because her parents thought her homeliness would preclude her being suitably married and wished to make her single life more tolerable with a good education. However, du Châtelet, grew into a beautiful young woman of intelligence and wit who not only married, but had a well-known series of lovers including a very long-standing relationship with Voltaire.[20] Her significant intellectual interests were in physics and mathematics.

In collaboration with Voltaire, she wrote *Eléments de la philosophie de Neuton* (1738) that explained Newtonian physics for a French audience. *Institutions de physique*, published in 1740, originated as a physics textbook for her son and included principles from Newton and German mathematician Gottfried Leibniz. By now, students were arriving to study with du Châtelet and she began the culmination of her life's work, a two-volume translation of Newton's *Principia* into French. It was published in 1759, ten years after her death, which occurred a few days after the birth of her fourth child [22, 24].

---

[18]Atomism was a natural philosophy that held that substances were comprised of indestructible atoms of infinite shape and size [27].

[19]The only female presence in almost 300 years at the Royal Society was a skeleton in the anatomy collection.

[20]Voltaire was a French philosopher and writer who helped define the movement called the Enlightenment. He advocated for freedom of speech and religion and separation of church and state [28, 29].

2.12   Laura Bassi: Physicist (1711–1778)

Laura Bassi, an Italian physicist, was fortunate to live in Bologna, Italy, a city that prided itself on being a leading center for women in education. At the age of 20, she was presented with membership in Bologna's Academy of Sciences, which was part of the Institute of Sciences. Shortly thereafter, she received a doctorate in philosophy from the University of Bologna. Although she was offered a chair in philosophy at the University and named a university professor, because she was a woman, she was allowed to give public lectures only by invitation. She was able to overcome significant resistance from various circles in Italian society as her career advanced due to the support of her husband as well as members of the academic community, church leaders, and political figures.

She published scientific papers as a result of some of her research. They involved topics including air pressure (1745), solutions for problems in hydraulics (1757), the use of mathematics to solve trajectory problems (1757), and bubbles formed from liquids in gas containers (1791). Much of the rest of her research, which did not result in publications, involved fluid mechanics, Newtonian physics, and electricity. Because of her social position in Bologna society, she was also required to write poetry for community events and contribute to literary publications [23].

2.13   Maria Gaetana Agnesi: Mathematician (1718–1799)

In 1748, Italian Maria Agnesi published a two-volume, 1020-page manual titled *Analytical Institutions* that significantly enhanced the mathematical and scientific knowledge of the day. The volumes, intended as a textbook for her younger brothers, included analysis of finite quantities (algebra and geometry) in volume one, and differential and integral calculus (analysis of variable quantities and their rates of change) in volume two. Her clarification of the work of the best known mathematicians and scientists of the day, including Leibniz, Newton, Kepler, Galileo, and L'Hopital, was recognized for its importance around the world and translations were sought by scientists and mathematicians. And all of this, from a woman born in Milan, Italy into a society where most young women, even in the upper classes, were not even taught to read.

The French Academy of Sciences described her work on infinitesimal analysis as "organized, clear, and precise" and authorized translation of her second volume from Italian into French in 1749. The English translation was published in 1801. In 1750, she was named honorary chair of mathematics and natural philosophy at the University of Bologna, although she never lectured. After her father's death in 1752, she gradually withdrew from mathematical and scientific activities, apparently because she associated those activities with him and the strong encouragement and support he had provided to her in her endeavors [22, 24].

2.14   Jane Colden: Botanist (1724–1766)

The first woman botanist in the U.S., by 1757 Jane Colden (later Farquhar) had prepared a catalog of over 300 local species of flora and had exchanged specimens and seeds with several colonial and European botanists. Under the tutelage of her father, Cadwallader Colden, a New York botanist and government leader, Jane Colden mastered the Linnaean classification system and wrote a paper for a publication of the Edinburgh Philosophical Society.[21] She is best known for her identification and description of the gardenia, which she was the first to identify. In conformance with the social mores of her time, her botanical work ceased after her marriage in 1759 [1, 12, 15, 20].

2.15   Caroline Herschel: Astronomer (1750–1848)

An astronomer in her own right, German Caroline Herschel (Fig. 2.8), sister of famed astronomer William Herschel, was the first woman to discover a comet. In early 1783, by which time she had moved to England to join William, she discovered the Andromeda and Cetus nebulae, and by the end of the year, had discovered an additional 14 nebulae. Between 1786 and 1797, Caroline Herschel discovered eight comets. These discoveries made her famous in the astronomical community and brought her the attention of the King of England who gave her a salary of 50 pounds per year to officially recognize her as William's assistant.

Caroline Herschel next undertook to reorganize the star catalogue of John Flamsteed, the first Astronomer Royal. *The Catalogue of Stars...* was published by the Royal Society in 1798 and contained an index to every observation of every star made by Flamsteed and a list of upwards of 560 stars that are not inserted in the British Catalogue. At the age of 75, Caroline Herschel finished a catalogue that recorded the positions of about 2500 nebulae for which the Royal Astronomical Society awarded her a gold medal. At age 85, she was made an honorary member of the Royal Astronomical Society and was similarly honored by the Royal Irish Academy. On her 96th birthday, she was awarded the Gold Medal of Science by the King of Prussia [1, 25].

---

[21] Cadwallader Colden encouraged his daughter to study botany as he thought this was a field particularly suited for women: "their natural curiosity and pleasure they take in the beauty and variety of dress seems to fit them for it."

Fig. 2.8 Caroline Herschel – Wikipedia

2.16   Marie Anne Pierrette Paulze Lavoisier: Chemist (1758–1836)

Antoine and Marie Lavoisier (Fig. 2.9) established chemistry as a modern scientific discipline. Their discoveries included the identification of oxygen and the nature of combustion, oxidation, and respiration. In addition, they established the law of conservation of matter as a principle for experimental design. It is impossible to separate Marie's contributions from Antoine's although she is known to have assisted with experiments and kept all of the laboratory records and notes. She edited and illustrated her husband's treatise *Elements of Chemistry* (1789). She translated scientific papers from English to French and provided commentaries on them as well. In particular, she translated, wrote commentaries on, and added correcting footnotes to Richard Kirwan's 1787 *Essay on Phlogiston and the Constitution of Acids* [1, 25].

Fig. 2.9 Marie Lavoisier and her Husband – Wikipedia

2.17   Jane Haldimand Marcet: Science Popularizer (1769–1858)

Briton Jane Marcet wrote books to popularize science – writing books and papers that explained science to general readers – specifically intended for women and young people. She is particularly remembered for the impact her *Conversations in Chemistry* had on influencing future scientist Michael Faraday (for whom the unit of electrical capacitance is named).[22] Jane was encouraged to begin a writing career by her husband, physician Dr. Alexander Marcet, whose passion for chemistry exceeded his interest in practicing as a physician. *Conversations in Chemistry* (1806) was very popular and went through numerous editions, including 15 American editions titled *Mrs. Bryan's Conversations*. Jane Marcet believed that the information presented in a conversational format was more readily comprehended by the audience, as she was better able to understand chemistry after conversing with a friend. Her other books included *Conversations on Botany, Conversations on Natural Philosophy, Conversations on Political Economy,* and *Conversations on Vegetable Physiology* [1, 2, 5, 25, 30].

2.18   Sophie Germain: Mathematician (1776–1831)

Frenchwoman Sophie Germain was so determined to study mathematics that she persevered even after her parents made sure her bedroom was without light or fire and that she was left without clothes so that she would have to stay in bed. She persevered after she was not allowed into the École Polytechnique at 18 years of age to continue her studies because she was a woman. Her parents had finally relented and allowed her to study mathematics during the day and now she was ready for more advanced education. Undeterred by the refusal of the École Polytechnique to admit her, she studied on her own through notes obtained from other students. She wrote to French mathematician Joseph Louis LaGrange under a pseudonym and he was so impressed with her comments, that he met with her and commended her observations. She later also communicated with German mathematician Carl Friedrich Gauss and he was so impressed that, in 1831, he was successful in having the University of Göttingen award her an honorary degree.

Vibrations and their patterns became the subject of a competition for the French Academy of Sciences. Germain's work, the only entry in the competition, was awarded the grand prize in 1816. The vibration patterns are used today in the construction of tall buildings, such as skyscrapers. Now that she had a prize, she was allowed to attend sessions of the Institut de France. In the 1820s, Germain became interested in number theory and developed a theorem in support of Fermat's Last Theorem, her most important work in number theory. Her theorem, referred to today as Germain's Theorem, has since been generalized and improved, but not replaced [24].

---

[22] Pioneering physicist Michael Faraday produced the first electrical generators and motors. The farad, the unit of electrical capacitance, was named for him He introduced the terms "electrolyte" and "ions" into our scientific vocabulary.

2.19 Mary Fairfax Somerville: Science Popularizer (1780–1872)

One of the first honorary women members of the Royal Astronomical Society (Great Britain), Briton Mary Somerville (Fig. 2.10), like Jane Haldimand Marcet, helped to popularize science particularly for women. After spending a year at a boarding school when she was 10 years old, Mary developed a thirst for reading and arithmetic. She taught herself Latin, and then algebra, after seeing strange symbols in a ladies' fashion magazine. When her parents found out about her interest in mathematics, her father forbade her study due to worries that mental activity would harm her female body. After her first husband died and left her with a modest inheritance, she openly educated herself in trigonometry and astronomy. Most of her friends and family did not support her educational efforts.

Fig. 2.10 Mary Fairfax Somerville. (Courtesy Library of Congress)

Mary married her first cousin, Dr. William Somerville, and found in him some-one to support her pursuits of educational and intellectual matters. In fact, William encouraged Mary to expand her studies beyond mathematics and astronomy to Greek, botany, and mineralogy. In 1834, she published *On the Connexion of the Physical Sciences* which presented a comprehensive picture of the latest research in the physical sciences. Her 1831 book, *Mechanism of the Heavens*, contributed to the modernization of English mathematics. Mary was occasionally criticized for her "unwomanly" pursuit of science, nevertheless, she was referred to, both in England and abroad, as "the premier scientific lady of the ages" [23].

2.20   Josephine Ettel Kablick (Josefina Kablíková): Botanist and Paleontologist (1787–1863)

An intrepid Czech botanist and paleontologist, Josephine Kablick collected plant and fossil samples. Undeterred by any weather or terrain, she gathered new species in dark forests and on mountains. Her collection gained renown and she gradually collected plants for schools and colleges in her country as well as for museums and learned societies in other parts of Europe. Fittingly, many of the fossils and plants that she collected are named in her honor [31].

2.21   Almira Hart Lincoln Phelps: Science Writer and Educator (1793–1884)

Almira Hart Lincoln became interested in science after the death of her husband in 1823 and her subsequent return to teaching on the staff of her sister's (Emma Hart Willard) school, the Troy Female Seminary. She was encouraged by Rensselaer professor Amos Eaton. In 1829, she published her first science textbook, *Familiar Lectures on Botany*. After she married John Phelps in 1831, she continued to write and revise her textbooks in addition to taking care of her family responsibilities. Other volumes included *Chemistry for Beginners* (1834) and *Familiar Lectures on Natural Philosophy* (1837).

She served as the principal for several female seminaries and emphasized science in the curriculum. Her teaching innovations included experimental methods in chemistry and botany. In 1859, she was the third woman elected as a member of the American Association for the Advancement of Science. Although she supported educational equality for women, she opposed suffrage and was active in the Woman's Anti-Suffrage Association. Upon her death, her herbarium of plants, col-lected throughout her lifetime, was presented to the Maryland Academy of Sciences, of which she was the first woman member [1, 20, 25, 32].

2.22   Margaretta Hare Morris (1797–1867)

One of the first women entomologists, Margaretta Hare Morris contributed to agricultural science through her studies of the 17-year locust and the Hessian fly. Morris discovered, contrary to the belief at the time, that Hessian fly larvae were of two separate species. She also discovered the main predator for the flies.

In 1840, she published her first paper on the subject which appeared in *Transactions of the American Philosophical Society.*[23] Morris published her discoveries of 17-year locusts demonstrating how the locusts preyed on trees first in 1846 and later in 1850. She is considered the first practicing woman naturalist and was recognized for her illustrations of botanical papers. She wrote articles for the *American Agriculturist* on topics ranging from moths, natural history and fleas, to bedbugs, moths, and the 17-year locust. Morris was the second woman elected to the Philadelphia Academy of Natural Sciences. In 1850, she and Maria Mitchell were the first women elected to the American Association for the Advancement of the Science [15, 25, 33].

2.23   Mary Ann Anning: Paleontologist (1799–1847)

Mary Ann Anning (Fig. 2.11) became a fossilist (a paleontologist) at age 12 when she discovered a fossil, sometimes called Mary's Monster or the Lyme Crocodile, that was in fact an *Ichthyosaurus*, a porpoise-like marine dinosaur from the Jurassic period. Found on the beach in Dorset County, England (near the town of Lyme Regis), this discovery, and the resulting sale of the bones to the British Museum, allowed Anning to support the family after her father died.

Anning continued to find new and unusual fossils including a *Plesiosaurus* skeleton (1823) and a flying reptile skeleton *Pterodactylus macronyx* (1828). Anning also discovered a *Squaloraja* fish fossil that established a link between modern sharks and rays. Although ignored by many scientists because of her gender, Anning was granted a small stipend in 1838 from the British Association for the Advancement of Science. She also received a small stipend from The Geological Society of London and was named the first honorary member of the Dorset County Museum [12, 25].

---

[23] In those days, papers were published in leading scientific publications after they were read to the membership (all-male). Thus, Morris needed to find men willing to read her papers. Because her work was well-respected, she had no trouble finding men willing to do so.

Fig. 2.11 Mary Ann Anning – Wikipedia

2.24   Ada Byron Lovelace: Mathematician and Computer Scientist (1815–1852)

The daughter of the English poet Lord George Byron, Ada Lovelace now has a computer language named (Ada) after her. A somewhat sickly child, Lovelace was tutored at home and was competent in mathematics, astronomy, Latin, and music by the age of 14. Totally enthralled by Charles Babbage's Difference Engine (an early computer concept), at 17 years old, Lovelace began studying differential equations. As proposed, his second machine, the analytical engine, could add, subtract, multi-

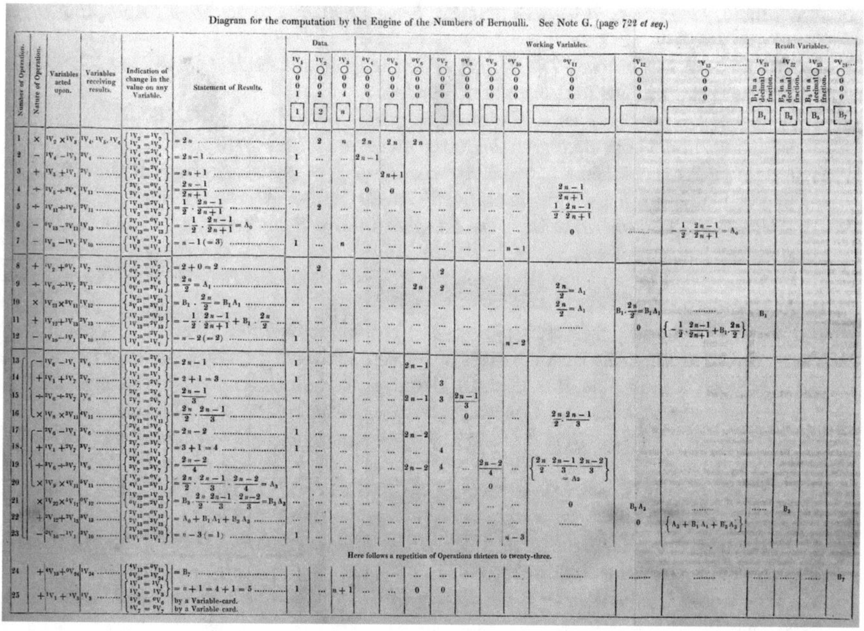

Diagram for the computation by the Engine of the Numbers of Bernoulli. See Note G. (page 722 *et seq.*)

Fig. 2.12 Ada Byron Lovelace – Note G – Wikipedia

ply, and divide directly and it would be programmed using punched cards, the same logical structure used by the first large-scale electronic digital computers in the twentieth century.

In 1842, the Italian engineer, L.F. Menabrea published a theoretical and practical description of Babbage's analytical engine. Lovelace translated this document adding "notes" in the translation. Her notes constitute about three times the length of the original document and, as explained by Babbage, the two documents together show "That the whole of the development and operations of analysis are now capable of being executed by machinery." These notes include a recognition that the engine could be told what analysis to perform and how to perform it – the basis of computer software. Her notes (Fig. 2.12) were published in 1843 in *Taylor's Scientific Memoirs* under her initials, because although she wanted credit for her work, it was considered undignified for aristocratic women to publish under their own names. Ada Lovelace is considered to be the first person to describe computer programming [22, 24].

References

1. M.B. Ogilvie, *Women in Science: Antiquity through the Nineteenth Century, a Biographical Dictionary with Annotated Bibliography* (MIT Press, Cambridge, 1993)
2. C.A. Ronan, *Science: Its History and Development among the World's Cultures* (The Hamlyn Publishing Group Limited, New York, 1982)

3. P. Whitfield, *Landmarks in Western Science: From Prehistory to the Atomic Age* (Routledge, New York, 1999)

4. R. Spangenburg, D.K. Moser, *The History of Science from the Ancient Greeks to the Scientific Revolution* (Facts on File, Inc., New York, 1993)

5. C. Suplee, *Milestones of Science* (National Geographic, Washington, DC, 2000)

6. Glasses. https://en.wikipedia.org/wiki/Glasses. Accessed 9 May 2020

7. History of Optics. https://en.wikipedia.org/wiki/History_of_optics. Accessed 9 May 2020

8. Galileo Galilei. https://en.wikipedia.org/wiki/Galileo_Galilei. Accessed 9 May 2020

9. J.A. Baer, *Women in American Law: The Struggle Toward Equality From the New Deal to the Present*, 2nd edn. (Homes & Meier, New York, 1996)

10. B. Harris, *Beyond her Sphere: Women in the Professions in American History* (Greenwood Press, Westport, 1978)

11. E. Flexner, E. Fitzpatrick, *Century of Struggle: The Women's Rights Movement in the United States*, Enlarged edn. (The Belknap Press of Harvard University, Cambridge, 1996)

12. S.A. Ambrose, K.L. Dunkle, B.B. Lazarus, I. Nair, D.A. Harkus, *Journeys of Women in Science and Engineering: No Universal Constants* (Temple University Press, Philadelphia, 1997)

13. H. Garza, *Barred from the Bar: A History of Women in the Legal Profession* (Franklin Watts, New York, 1996)

14. E.A. Dexter, *Career Women of America: 1776–1840, Francetown* (Marshall Jones Company, New Hampshire, 1950)

15. M.W. Rossiter, *Women Scientists in America: Struggles and Strategies to 1940* (The Johns Hopkins University Press, Baltimore, 1992)

16. Merriam-Webster, Inc, *Webster's Dictionary of American Women* (Smithmark Publishers, New York, 1996)

17. Natural philosophy. https://en.wikipedia.org/wiki/Natural_philosophy. Accessed 9 May 2020

18. "A Brief History", The Vassar Planet – Backgrounders. www.Vassar.edu/relations/hist.html. Accessed 1 Jan 2002

19. Turner, Edna May, "Education of Women for Engineering in the United States 1885–1952," (Dissertation, New York University, 1954), Ann Arbor: UMI Dissertation Services

20. M.J. Bailey, *American Women in Science: A Biographical Dictionary* (ABC-CLIO, Denver, 1994)

21. "The Land Grant System of Education in the United States". http://www.ag.ohio.state.edu/~ohioline/lines/lgrant.html. Accessed 11 Apr 2000

22. M. Alic, *Hypatia's Heritage: A History of Women in Science from Antiquity through the Nineteenth Century* (Beacon Press, Boston, 1986)

23. Benjamin F. Shearer, B. S. Shearer (eds.), *Notable Women in the Physical Sciences: A Biographical Dictionary* (Greenwood Press, Westport, 1997)

24. C. Morrow, T. Perl (eds.), *Notable Women in Mathematics: A Biographical Dictionary* (Greenwood Press, Westport, 1998)

25. P. Proffitt (ed.), *Notable Women Scientists* (The Gale Group, Detroit, 1999)

26. B. F. Shearer, B. S. Shearer (eds.), *Notable Women in the Life Sciences* (Greenwood Press, Westport, 1996)

27. Atomism. https://en.wikipedia.org/wiki/Atomism. Accessed 9 May 2020

28. Voltaire. https://en.wikipedia.org/wiki/Voltaire. Accessed 9 May 2020

29. Voltaire, Stanford Encyclopedia of Philosophy. https://plato.stanford.edu/entries/voltaire/, revised 20 July 2015

30. J. Daintith, D. Gjertsen, *A Dictionary of Scientists* (Oxford University Press, Oxford, 1999)

31. H.J. Mozans, *Woman in Science with an Introductory Chapter on Woman's Long Struggle for Things of the Mind* (D. Appleton and company, New York, 1913). Converted to a kindle edition by John Augustine Kahm

32. L.B. Arnold, *Four Lives in Science: Women's Education in the Nineteenth Century* (Schocken Books, New York, 1984)

33. "Margaretta Hare Morris", Pennsylvania Conservation Heritage Project. https://paconservationheritage.org/stories/margaretta-hare-morris/. Accessed 29 Mar 2020

Chapter 3
The Fight for Educational Equality

3.1 Introduction

The availability of secondary and then higher education for women in the U.S. by the end of the nineteenth century, the highest level in the world, led to women's greater entry into the scientific professions for two primary reasons. First, the higher level of education allowed more women to study science systematically. Second, the higher levels of education led to greater numbers of jobs and increasing numbers of professorships becoming available to generations of women in science. By the late nineteenth century, enough science professorships existed at the women's colleges to enable the formation of a nucleus of a female enclave within the American scientific community [1].

3.2 Women's Colleges

The four decades after the Civil War saw a dramatic expansion in the opportunities for higher education for women. This expansion has been attributed to (1) ideological pressure from the women's suffrage movement, and (2) the need for enrollment at colleges and universities due to dropping enrollments first to the manpower requirements of the Civil War, then to economic depression, and subsequently due to dissatisfaction with the college curriculum. In the West, most institutions of higher learning were state-supported and coeducational from the start. However, their women students were primarily educated for marriage and appropriate female jobs, such as public school teaching [2].

© The Editor(s) (if applicable) and The Author(s), under exclusive license to
Springer Nature Switzerland AG 2020
J. S. Tietjen, *Scientific Women*, Women in Engineering and Science,
https://doi.org/10.1007/978-3-030-51445-7_3

The first women's colleges were the Georgia Female College (1836) and Elmira (New York), which opened in 1855.[1] The Georgia Female College was the first institution of any sort to grant degrees to women in the U.S. Among its required courses were chemistry, mineralogy, and astronomy. The foundation of Vassar College in 1861, however, heralded a new era in women's education. Matthew Vassar wrote, "It occurred to me that women, having received from her creator the same intellectual constitution as man, has the same right as man to intellectual culture and development." His institution would "accomplish for young women what colleges of the first class accomplish for young men: that is, to furnish them the means of a thorough, well-proportioned, and liberal education, but one adapted to their wants in life." Most of the women who attended the women's colleges were interested in education for its own sake or in preparation for teaching careers. Other women's colleges followed Vassar: Smith and Wellesley in 1875, Bryn Mawr in 1885, Mount Holyoke which began granting degrees in 1888, and Radcliffe in 1893[2] [2–4].

Some coeducational private universities were established during this period as well. Saint Lawrence University, Boston University, and Swarthmore date from the 1860s and 1870s. Cornell opened its doors to women in 1875, although it had previously been a men's college [2]. Although the Massachusetts Institute of Technology (MIT) did not officially admit women until 1876, Ellen Swallow (later Richards) had been accepted as a "special student" in the Fall of 1870 and allowed to study without paying tuition. She graduated with a BS in chemistry in 1873 and thus became the first woman in the U.S. to enter a strictly professional scientific school [5].

The women's colleges were the most receptive to hiring female professors. Maria Mitchell (Fig. 3.1), an outstanding astronomer, became one of the first female members of the Vassar faculty, as chair of astronomy and as the first director of the observatory [2, 5]. Many of the first generation of women faculty, like Maria Mitchell, did not have a strong educational background. Most had attended a female seminary or academy, some had graduated from other early colleges, but many had little formal training. By the early 1900s, however, doctorates became a requirement for faculty at the more highly ranked women's colleges. In keeping with the social mores of the time, the women faculty could not be married. Conversely, with the exception of Bryn Mawr College, unmarried men could not be hired as faculty! [1].

The women's colleges played a crucial role in the careers of academic women. Of the 504 women listed in the first three editions of *American Men of Science*, almost 40% had attended eight of the eastern women's colleges and 57% of all women listed worked at the women's colleges[3] [1, 2]. Most of the early women

---

[1] Elmira College is today a private, coeducational, liberal arts college.

[2] One of the major problems of the early women's colleges was finding enough adequately qualified students. Wellesley ran a preparatory school until 1880. Vassar ran theirs until 1888.

[3] The first three editions of James McKenn Cattell's *American Men of Science: A Biographical Dictionary* (1906, 1910, and 1921) listed 504 women. Of a group of 483 of them, almost 40% had attended eight eastern women's colleges. The women's colleges were major employers of female scientists. Between 1900 and 1920, they employed 21 of the 23 female physicists listed by Cattell.

Fig. 3.1 Maria Mitchell.
(Courtesy of Library of
Congress)

356   HARPER'S NEW MO]

MISS MARIA MITCHELL, PROFESSOR OF ASTRONOMY.

science faculty had little time to publish and therefore to advance their scientific careers, however. They were busy with heavy teaching loads (often in more than one science), planning the college's first science buildings including selection of laboratory apparatus, and finding protégés to succeed them. Later generations of women science faculty at the women's colleges, usually armed with a doctorate, could do research and teach their specialty within the department, if they could find adequate financial resources to do so. The lack of adequate financial resources and associated personnel as well as the belief that a segregated experience at the graduate level was not desirable were the main reasons that the women's colleges did not in general become women's universities with graduate schools and offer the desired doctorate themselves [1].

In 1906, 57% of all the women whom he listed worked at these institutions. By 1921, the percentage had fallen to 36%, although the number of jobs doubled, primarily because opportunities were opening up at the larger state and private universities.

3.3   The Doctorate

When American graduate schools were founded, women quickly learned that the doors were not open to them. These schools were patterned on the German universities that had established the Doctor of Philosophy programs in the eighteenth century and had never admitted women. Successes in opening those doors in the U.S. came as early as the 1870s at Boston University and as late as the 1960s at Princeton University. Most of the skirmishes in this battle occurred during the decades of the 1880s through the 1900s [1].

When women did apply to graduate school after the Civil War, the standard reply was that there was "no precedent" for admitting a woman, regardless of her qualifications, and regardless if she were the first or the nth – no precedent for admitting had been set and no one seemed willing to set one. When Ellen Swallow (later Richards) applied to MIT for a graduate degree in chemistry in 1870, she was turned down solely on the grounds that MIT did not want its first graduate degree to go to a woman. Yet, she was admitted as a candidate for a second bachelor's degree and as a "special student," without having to pay tuition, so that MIT could deny later that she was officially enrolled. Johns Hopkins admitted Christine Ladd (Fig. 3.2) as a "special student" in 1878 but her name did not appear in the catalog and her admittance thus did not set a precedent.[4] After she completed her PhD program in 1882, she was told that a degree would not be awarded. Mary Whiton Calkins was told the same thing at Harvard University after passing her exams in an exceptional manner in 1895. Ladd-Franklin did finally receive her earned PhD in 1926, only 42 years later! Calkins later refused to accept the Radcliffe PhD offered her believing she merited her earned Harvard PhD. Such was not to be[5] [1, 2].

Some smaller institutions found a way to award doctorates to women starting as early as 1877 and continuing into the 1880s, and some of these were in the sciences. The institutions included Syracuse University, Boston University, University of Wooster, Smith College, the University of Michigan, and Cornell University. Oftentimes an enthusiastic chancellor or president was championing graduate education and would welcome women as students [1].

Articles and books began to appear after 1889 surveying the graduate education situation in the U.S. and abroad. Then in the early 1890s, six institutions decided to admit women to graduate school on the same basis as men. In 1891, Stanford and the University of Chicago (both western institutions) announced that full coeducation

---

[4] Johns Hopkins was proclaimed as the first full-scale graduate school in America when it opened in 1876. Apparently, its trustees were not sympathetic to women and its first President, Daniel C. Gilman, felt he had enough issues to deal with without having to deal also with coeducation.

[5] In 1895, Mary Whiton Calkins passed all of her examinations at Harvard but was denied her PhD. In 1898, Ethel Puffer (later Howes) passed all of her examinations with such distinction that a committee of eight full professors deemed her to be unusually qualified for a Harvard PhD but it was not to be granted. In 1902, Harvard established the Radcliffe Graduate School to grant university degrees to women and both Puffer and Calkins were offered Radcliffe PhDs; Puffer accepted but Calkins refused insisting on a Harvard degree or none at all.

Fig. 3.2 Christine
Ladd-Franklin. (Courtesy
of Library of Congress)

was being introduced at the graduate and undergraduate level. Columbia and Brown
tied their decisions to admit women to graduate school to the formation of a coordinate college for women undergraduates. Yale and the University of Pennsylvania
both decided to admit women to graduate school but continued to deny them admittance to undergraduate school.[6] Brown University established a separate and distinct Women's College as early as 1886 through recommendation of the Corporation,
but President Andrews actually admitted women in 1891 [1, 6, 7].

However, many major American graduate schools still continued to exclude
women as did the German universities. Women who had been working to surmount
this issue decided in 1888 that a graduate fellowship awarded to a woman who
wished to study abroad might convince recalcitrant universities of the caliber of

---

[6]Yale and the University of Pennsylvania admitted women in 1892. Barnard College, a coordinate
college for women, was established by Columbia University in 1889. It is named for Frederick
A.P. Barnard, the tenth president of Columbia College, who argued unsuccessfully for the admission of women to Columbia University.

women and the graduate educational opportunities available elsewhere. The first fellowship of the Association of Collegiate Alumnae (later the American Association of University Women) was awarded in 1890. The University of Göttingen became the first German university to award women doctorates in 1895, although those women were primarily not German nationals.[7] Johns Hopkins awarded its first doctorate to geologist Florence Bascom (but not retroactively to Ladd-Franklin) in 1893 as a "special student." Women were not admitted to all of its graduate programs officially until 1907. Harvard held out even longer, and began to award women doctorates only in 1963 [1].

By the early 1900s, a doctorate (a Doctor of Philosophy – PhD) in the sciences was a required credential to teach at colleges and universities, even the women's colleges. Because it was so difficult for women to obtain a PhD, particularly in the sciences, this was a virulent form of discrimination and one that segregated women from the mainstream of the scientific profession.

3.4   Professionalization

During the period 1820 through 1920, two major trends were converging in the U.S. First was the rise of educational levels of people in general and women in particular, including the rise of higher education and expanded employment for middle-class women. The second trend was the growth, bureaucratization, and professionalization of science and technology in the U.S. And because women were supposed to be soft, delicate, emotional, and noncompetitive, and science was seen as almost diametrically opposite – tough, rigorous, rational, and unemotional – a woman scientist was an oxymoron. As scientists, women were atypical women; and as women they were unusual scientists [1]. And the field of science in the U.S. was still not quite ready for these very accomplished women scientists.

The scientific societies rarely welcomed women. Some women did manage to get admitted to the scientific societies even as early as the mid 1800s. Lucy Way Say, the widow and former assistant of entomologist Thomas Say, was elected an associate member of the Academy of Natural Science in Philadelphia in 1841.[8] However, she did not attend the meetings. Maria Mitchell was elected the first (and for a long time, only) woman member of the American Academy of Arts and

---

[7] The professors at German universities were quite opposed to admitting German women for doctorates. Foreign women were less of a threat since they would return home and not expect to teach in Germany. As early as 1902 and as late as 1908, the German universities began admitting their own countrywomen and awarding them degrees. Some of the first German graduates were faculty daughters who had been watching the American women and waiting for their own chance to attend the universities and earn degrees.

[8] Thomas Say is considered the father of American entomology. Lucy Way Say was an illustrator for her husband's book and publications. She was the first woman member of the Academy of Natural Sciences of Philadelphia (her husband and his brother were co-founders and leaders). She may have been the first woman to join an American scientific society.

Sciences in Boston in 1848. She became a member of the American Association for the Advancement of Science (AAAS) in 1850 along with entomologist Margaretta Morris. Almira H. L. Phelps would become a member of AAAS in 1859. In 1869, Maria Mitchell became one of the first women in the American Philosophical Society of Philadelphia [1, 8–10].

Many social taboos on women's participation in public groups weakened during the Civil War. Thus, in the 1870s and thereafter, educated women wanted to join scientific societies and sought work in museums and observatories. Men appeared to be threatened by this supposed encroachment into what they perceived as their territory and a backlash began. The AAAS introduced the "fellow" level of membership in 1873 for those members "as are professionally engaged in science, or have by their labors aided in advancing science." Any existing members could be made a fellow but new members chosen as fellows had to go through a nominating committee and selection process. In 1874, despite her sex, Maria Mitchell was elected a fellow of AAAS [1, 8, 10].

In the 1880s and 1890s, the result of the backlash to more women entering the fields of science was women's almost total ouster from major or even visible positions in science. There was very strong resistance to women holding the same jobs as men in science, particularly university teaching positions or government employment. This resistance tended to be implemented through standardization of credentials, especially a requirement of a PhD for university faculty positions. The female faculty at the women's colleges now faced an ironic situation – they were no longer considered employable if they did not have their PhD. And since the primary institutions that offered doctorates – American and German universities – did not admit women to their graduate programs, women faculty with the now "necessary" qualifications became harder to find [1].

At the same time, the scientific societies were in the process of upgrading themselves to "professional" societies, thereby excluding individuals perceived as "amateurs." Thus, women, who had been precluded from important positions in science, if allowed to join, would diminish the prestige associated with belonging to such a professional society. Some societies restricted women to lower levels of membership or required women to have more degrees for admission. Some women turned to women's clubs in science or a separate women's scientific society. Neither of these alternatives was a suitable substitute for admission into the mainstream professional scientific societies[9] [1].

The American Ornithological Union was established in 1883 as all male. When a woman applied for membership in 1885, she was accepted as an "associate member." The American Chemical Society, established in 1874 and electing at that time a woman as one of the interim committee's honorary vice presidents, lost all of its

---

[9]There were many local all-women science clubs including the Female Botanical Society of Wilmington, Delaware in the 1840s; the Dana Natural History Society of Albany, New York in the 1860s, the Syracuse Botanical Club, and the Botanical Society of Philadelphia. Other clubs existed in Jersey City, New Jersey in the 1870s and 1880s and in Boston in the 1890s and early 1900s. The (Women's) Natural Science Club was established in Washington in 1891.

female members by 1880 and did not have any women as members again until 1891. The all-male Anthropological Society of Washington in 1885 refused admittance to Matilda Coxe Stevenson. She then met with other women and formed the Women's Anthropological Society of Washington in 1885. Stevenson was finally admitted to membership in the Anthropological Society in 1891 and the two groups merged in 1899. These organizations were eclipsed by the founding of the American Anthropological Association in 1902 that finally elected its first woman president in 1940 [1].

The American Society of Naturalists was established in 1883 with criteria requiring that potential members be "professionally engaged."[10] Its first woman member was elected in 1886 followed by a second woman in 1892. These women, Emily Gregory and Ida Keller, appear to have been acceptable as members because they both had foreign doctorates. Apparently the Geological Society of America, established in 1888, required a doctorate for women, but not for men, to be members, although this requirement was never stated in its regulations [1, 8].

The Botanical Society of America, whose leaders split off from the Torrey Botanical Club when that organization's membership reached about 40% women, was established in 1893. Membership was at first restricted to those botanists who had published several articles or otherwise made important "professional" contributions to science. Only one woman was among its 25 charter members, Elizabeth Knight Britton, one of the most prominent women scientists of the 1890s. Britton published over 300 articles on mosses in her lifetime. However, she did acquiesce to the group's unwritten prohibition on female attendance, "I think it will not be a meeting suitable for women to attend, as the discussions will be rather embarrassing and the men will probably prefer to dine alone" [1, 8, 10].

Some women may have been accepted as members at the founding of some scientific societies in the 1880s and 1890s; but these women were the truly exceptional women who already had the best credentials, the best positions, the most publications, or personal relationships with prominent men in the field. Although no women were members of the American Psychological Society when it was founded in 1892, one woman became a member in 1894, another was elected to its council in 1897, and the first woman, Mary Whiton Calkins, was elected president in 1905. The American Mathematical Society had two women among its charter members in 1894 and a woman on its council and serving as vice president by 1906. Two women were members at the founding of the American Physical Society in 1899. The Association of American Geographers, established in 1903 before any American

---

[10]Membership requirements of the American Society of Naturalists: *Membership in this society shall be limited to persons professionally engaged in some branch of Natural History as Instructors in Natural History, Officers in Museums and other Scientific Institutions, Physicians, and others. The following persons shall be considered professionally engaged in natural history within the meaning of Article II, Section 1: – Only those who regularly devote a considerable portion of their time to the advancement of natural history; first, those who have published investigations in pure science of acknowledged merit; second, teachers of natural history, officers of museums of natural history, physicians, and others who have essentially promoted the natural-history sciences by original contributions of any kind.*

university had a doctorate in the field, required its female members, but not its male members, to have doctorates, which meant they must have had foreign training [1].

By the early 1900s other women had begun to challenge the de facto discrimination against women members attending meetings and especially banquets of the scientific societies. Mary Whitney of Vassar College who had attended the founding meeting of the American Astronomical Society in 1899 was not sure she and her protégé Caroline Furness would be welcome at the organization's banquet in 1902. The President wrote her to assure her that the women would be welcome "I am much disappointed to notice that although you hope to be here at our meeting, you do not propose to join in the dinner. Possibly you may be under a misapprehension, supposing that the dinner is only for the men of the society. Permit me, therefore, to assure you that all members are equal, and that we should like very much to have our lady members with us" [1, 8, 10].

Smokers and the habit of smoking also served to discriminate against women scientists. Properly bred women did not enter rooms where men were smoking or smoke themselves until into the 1920s. Most women would not attend a smoker unless men specifically requested their presence. Thus, Sarah Whiting began to attend the banquets and smokers of the American Physical Society between 1907 and 1909 when so encouraged by the organization's President. Many other organizations used the smoker to actively discourage women from attending as did the Association of American Geographers as late as 1915 [1, 8, 10].

Although not a professional society, the publication *American Men of Science,* which premiered in 1906, was a biographical reference on the leading scientists in America. Certainly, the name by itself indicates the evident bias at the time of its first publication.[11] Through at least 1950, it was the chief source of numerical and biographical data on American scientists who had as its publisher said "carried on research in science." Entries that were starred indicated eminence in one's profession. It was important for scientists to be listed in AMS and to be starred [1, 11].

Women gradually, but slowly, won acceptance in scientific societies. It was not until after World War I that the National Academy of Sciences elected its first female member, Florence Sabin, in 1925. Its second female member, Margaret Washburn, was elected in 1931 [2, 8, 10].

3.5   Early Twentieth Century

Although there were quite a number of women in science by the early 1900s, few of the even excellently qualified women were in the most prominent or visible places. Before 1910, the women scientists did not seem to realize the second-class citizen position to which they had been relegated, since many seemed to still be celebrating the successes accomplished since the 1870s. When quantitative evidence appeared

[11] In 1971, the name was changed to *American Men and Women of Science.*

in 1910 that showed how far women had to climb to reach full equality in science, feminists began to understand that "women's work" in science would not lead to equality and that popular attitudes and cultural expectations needed to be attacked head on [1].

Not a single female scientist held a job in industry prior to World War I. Male scientists could choose between jobs in academia and industry. And in academia, they had many more choices as well. Men worked in universities and medical schools. Women were concentrated in lower positions at women's colleges, nursing and normal schools, museums, and research institutions. Men's and coeducational colleges and universities rarely hired women faculty members. Women scientists were often assigned to the home economics department instead of the department of their closer field, i.e., biology, chemistry or physics. At coeducational colleges, unmarried women faculty were placed in charge of women's dormitories. Of course, no one would hire married women scientists. Some women faculty were not listed in faculty rosters in order to avoid confrontation and controversy with groups including boards of trustees, alumni, and other faculty [2, 8].

By 1921, women held 0.001% of professorships at male colleges and 4% of the professorships at coeducational institutions. However, women accounted for 68% of the professorships at women's colleges. It would take the women's movement, the fight for suffrage during the early twentieth century, and two World Wars to improve women's position in the sciences [1, 2].

***

KEY WOMEN OF THIS PERIOD

***

3.6   Maria Mitchell: Astronomer (1818–1889)

In my younger days, when I was pained by the half-educated, loose, and inaccurate ways which we [women] all had, I used to say, "How much women need exact science," but since I have known some workers in science who were not always true to the teachings of nature, who have loved self more than science, I have now said, "How much science needs women" [1].

One of the most famous early American women scientists was Maria Mitchell (Fig. 3.1), an astronomer who discovered a comet in 1847. A founding faculty member at Vassar College and director of its observatory, Mitchell inspired a generation of women astronomers and scientists.

The third of ten children in a Quaker family, Mitchell's parents valued education for their daughters as well as their son. In addition, her father was an amateur astronomer and he taught her to use telescopes and navigational instruments. By the age of 12, Maria was assisting her father with astronomical observations and calculations of positions and orbits of celestial bodies as well as the timing of solar eclipses. However, since observatories did not hire women, Maria became a teacher and a librarian on Nantucket Island in Massachusetts.

In 1831, the King of Denmark offered a gold medal to the first person who discovered a comet visible only with a telescope. On October 1, 1847, Maria Mitchell sighted such a comet and calculated its exact position. Although observers in Europe found the comet within a few days, Mitchell's priority was determined within a year and she was awarded the prize. As a result of the prize, she became quite visible and was nominated and elected unanimously as the first woman member of the American Association for the Advancement of Science (AAAS) in 1850, becoming a fellow in 1874. In 1848, she became the first woman elected to the American Academy of Arts and Sciences.[12] In 1857, she chaperoned a daughter of a Chicago banker on a trip to Europe and was able to meet with Caroline Herschel and Mary Somerville.

When Matthew Vassar invited her to be on the faculty on his new college, Mitchell and her father moved to Poughkeepsie, New York, and she became the director of the observatory there as well as a faculty member. The Women's Educational Association of Boston raised the money to purchase her one of the best telescopes in the country. Among the Vassar students who were inspired by Maria Mitchell were Ellen Swallow Richards and Christine Ladd-Franklin.

In 1868, Mitchell made the first daily photographic series of sun spots. She also studied binary or double stars and published her observations of Jupiter and Saturn and their satellites. She wrote some popular science articles and edited the astronomy section of *Scientific American*.

In addition to astronomy, Mitchell turned her energies to expanding educational opportunities for women including opportunities in science. She helped establish the Association for the Advancement of Women, serving as president for two years and chairing its Committee on Science until her death.

Mitchell received two honorary LLD degrees, one from Hanover College in Indiana (1853) and one from Columbia University (1887). She also received an honorary PhD from Rutgers Female College (1870). A crater on the moon is named after her. The Maria Mitchell Association on Nantucket worked to improve education on Nantucket and to encourage women astronomers, providing them fellowships to study at the Harvard College Observatory. Maria Mitchell has been inducted into the National Women's Hall of Fame [8–10, 12].

3.7 Emily Lovira Gregory: Botanist (1840–1897)

In 1886, Emily L. Gregory, a botanist, was the first woman elected to the American Society of Naturalists, presumably due to her doctorate from the University of Zurich. She taught at Barnard College after brief stints at Bryn Mawr and the

---

[12] Botanist Asa Gray so objected to Maria Mitchell's election that he erased Fellow from her certificate and wrote honorary member instead. No other woman would be elected to the Academy until 1943.

University of Pennsylvania as an unpaid professor because she had an independent income. She went to Europe during four summers to buy the equipment that she needed for her laboratory at Barnard. This equipment, which she purchased using her own funds, included such items as microscopes, books and models. Her textbook, *Elements of Plant Anatomy*, was published in 1895. In 1899, the Barnard Botanical Club dedicated a bronze tablet in her honor as the first professor of botany at Barnard [2, 8].

3.8   Ellen Henrietta Swallow Richards: Chemist (1842–1911)

Ellen Swallow (Fig. 3.3) was admitted to MIT as a "special student" and earned a bachelor's degree there in chemistry in 1873 (the first woman graduate of MIT) after having graduated from Vassar (as one of its first graduates) in 1870. She was denied an earned doctoral degree from MIT as the school did not want a woman to be the first person awarded a doctorate in chemistry; she would have been the first student to have earned such a degree. While she was a graduate student, Swallow

Fig. 3.3 Ellen Henrietta
Swallow Richards –
Library of Congress

executed a complete survey of Massachusetts drinking water and sewage for the Massachusetts Board of Health (1872), taking more than 40,000 samples. Through this work, she warned of early inland water pollution. She also contributed the first Water Purity Tables and the first state water quality standards in the U.S. From 1873 to 1878, she taught in the MIT chemistry department without a title or salary as the first women teacher. She also did extensive research in mineral analysis.

After her marriage in 1875 to Professor Robert H. Richards, head of the department of mining engineering at MIT, Richards persuaded the Women's Education Association of Boston to contribute the funds needed for the opening of a Woman's Laboratory at MIT. As assistant director to Professor John M. Ordway, an industrial chemist, in the laboratory, Richards began her work of encouraging women to enter the sciences and to provide scientific training to women. In 1879, she became the first woman member of the American Institute of Mining Engineers. She was certainly technically qualified for this membership classification; however, her husband's status of vice president of the organization contributed significantly to her selection.

By 1883, the Women's Laboratory had proven so successful that MIT allowed women to enroll in regular classes and closed the laboratory. Richards' work in the laboratory had resulted in several books and pamphlets including the seminal *Food Materials and Their Adulterations* (1885). This publication influenced the passage of the first Pure Food and Drug Act in Massachusetts. Her work included analysis of air, water, and food and led to national public health standards and the new disciplines of sanitary engineering and nutrition. The interaction between people and their environment, her areas of study, have led to Richards being called the founder of the science of ecology.

In 1884, she was instrumental in setting up the world's first laboratory for studying sanitary chemistry. She served as assistant to Professor William R. Nichols in the new laboratory and held the post of instructor on the MIT faculty for the rest of her life. From 1887 to 1889 she supervised a highly influential survey of Massachusetts inland waters.

Since 1876, Richards had been on the forefront of promoting education for women, especially in science. In 1881, Richards helped found the Association of Collegiate Alumnae (later renamed the American Association of University Women). In 1882, she helped to organize the science section of the Society to Encourage Studies at Home.

After 1890, she concentrated most of her efforts on founding and promoting the home economics movement (at first it was called domestic science)—an achievement for which she is primarily noted (and frequently criticized for its detrimental effect on women's equality). Home economics was given definition by a series of conferences held in Lake Placid, New York, organized and chaired by Richards starting in 1899. She was involved in the formation of the American Home Economics Association (1908) and was appointed in 1910 to the National Education Association.

Richards' name is starred in early editions of *American Men of Science*. Smith College awarded Richards an honorary doctorate in 1910. She has been inducted into the National Women's Hall of Fame [1, 8, 9, 12–17].

3.9   Christine Ladd-Franklin: Psychologist (1847–1930)

"There is certainly a crying need for all the psychologists who have any logic in them to pull well together and to put up a good fight against all the irrational cranks!… Is this then a good time, my dear Professor Titchener, for you to hold to the mediaeval attitude of not admitting me to your coming psychological conference in New York – at my very door? So unconscientious, so immortal – worse than that – so unscientific!"

Christine Ladd-Franklin (Fig. 3.2) was one of the education pioneers of the late nineteenth century who laid the groundwork for women's graduate education. Her main goal was to prod universities to open their doors to women for graduate training and for women to be granted the degrees that they had earned. She was strongly motivated by her personal experience at Johns Hopkins University where she had completed the requirements for a doctoral program in logic and mathematics, but was denied the degree.

Christine Ladd attended Vassar College where she studied mathematics. Her favorite professor at Vassar was Maria Mitchell. After receiving her AB at Vassar in 1869, she spent nine years teaching secondary school. During this time, she wrote articles on mathematics for the *Educational Times*, an English publication. In 1878, she was accepted into the graduate program at Johns Hopkins University as a "special student."[13] After she successfully completed all of the requirements for a doctoral degree in logic and mathematics including a PhD thesis titled "The Algebra of Logic," school officials refused to award her the doctorate. It was finally awarded to her in 1926.[14] Ladd-Franklin was the first American woman to write on the topic of logic and became recognized for her contributions to symbolic logic.

Ladd-Franklin turned to the investigation of vision in the 1880s and began publishing articles on the subject in 1887. Her paper on vision, accompanied by several reviews of other people's vision research, was published in the first issue of the *American Journal of Psychology* and was the first American work in psychology by a woman. She went to Europe in 1891 and 1892 and performed experiments on color perception while in Göttingen. She attended lectures in Berlin and performed experiments there on the theory of color vision. Ladd-Franklin independently synthesized these ideas into her own color theory (the Ladd-Franklin color theory), which was presented to the International Congress of Psychology in London in 1892 in a paper titled "A New Theory of Light Sensation."

---

[13] Ladd-Franklin studied mathematics instead of physics because university laboratories did not admit women and mathematics did not require laboratory work.

[14] In 1926, Johns Hopkins righted what is called a "grievous wrong" and awarded Ladd-Franklin her PhD.

Upon her return to the U.S., Ladd-Franklin lectured at Johns Hopkins from 1904 to 1909. In 1914, she became a lecturer at Columbia, where she remained until 1927, although she was not paid a salary. Her name is starred in the first four editions of *American Men of Science*.

An avid advocate for women's educational progress, Ladd-Franklin also contributed her energy and money to increase opportunities for graduate study and faculty positions for women. She was active in the forerunner of the American Association of University Women and proposed the fellowship for study overseas that the organization established in 1890. From 1900 to 1917, Ladd-Franklin administered the Sarah Berliner fellowship that supported new women doctorates in research [1, 8, 10, 14, 18].

3.10   Sarah Frances Whiting: Physicist and Astronomer (1847–1927)

Sarah Whiting established the departments of physics and astronomy at Wellesley College and was probably the first person to introduce laboratory experiments for women. She trained a generation of women for careers in these emerging fields of science during her forty years at Wellesley. She developed the second laboratory physics course in the country and the first at a women's college. She also built and directed the Wellesley College Observatory.

Encouraged by her father, for whom she helped develop experimental demonstrations for his physics classes, Whiting graduated from Ingham University for Women in LeRoy, New York and then returned to teach mathematics and the classics. Wellesley College was being established at the time and Whiting was recruited by its founder, Henry Fowle Durant, to teach physics. She took courses at MIT and was appointed professor of physics at Wellesley in 1876. She observed the use of instruments for studying star spectra at the Harvard College Observatory and then introduced astronomy courses at Wellesley (originally called "applied physics"). She retired from teaching in 1912 but retained her directorship of the Wellesley College Observatory until 1916.

Whiting was awarded an honorary degree from Tufts University in 1905 and was elected a fellow of the AAAS in 1883. Whiting was a teacher, rather than a researcher, and one of her outstanding students was Annie Jump Cannon [1, 8, 10].

3.11   Mary Whitney: Astronomer (1847–1921)

Mary Whitney was an astronomer who trained under Maria Mitchell at Vassar College and graduated in the second class in 1868. After teaching for two years, Whitney studied mathematics and celestial mechanics at the University of Zurich. Unable to secure a university position, she taught school again when she returned from Europe. Whitney became Mitchell's assistant in the observatory in 1881 and succeeded her as director and professor of astronomy in 1888. Whitney hired Caroline Furness in 1894 as her assistant.

Whitney took on the responsibility of establishing scientific work at Vassar. Her research involved observations of comets, asteroids, double stars and variable stars. The staff at Vassar's observatory produced 102 publications during Whitney's tenure as director which was a huge accomplishment for a small college. Whitney is listed in the first three editions of *American Men of Science*. Whitney also worked for the advancement of women's educational and professional opportunities. She was the first President of the Vassar Alumnae Association. Active on the science committee of the Association for the Advancement of Women, she was a charter member of the American Astronomical Society. She also served as president of the Maria Mitchell Association of Nantucket which strove to improve science education [8, 10].

3.12   Matilda Coxe Stevenson: Ethnologist (1849–1915)

Matilda Coxe Evans Stevenson was an early ethnologist. She accompanied her husband James Stevenson, who was appointed an Executive Officer of the U.S. Geological Survey in 1879, on his research expeditions to the American West. They spent six months at Zuni Pueblo and its people and their lives became the focus of her research. She contributed significant research on all phases of pueblo culture and was the first person to concentrate on data on domestic and womanly matters. She made valuable studies of the child life of the Zuni, including their customs, habits, games, and ordinary activities.

Stevenson played a key role in opening opportunities for women anthropologists and pioneered in the use of observation and the gathering of data in the field. Stevenson's many reports were published in the *Annual Reports* of the Bureau of American Ethnology, at the Smithsonian Institution. She was the founder and first president of the Women's Anthropological Society in 1885. Later, she was elected a fellow of the American Association for the Advancement of Science. Her name is starred in the first two editions of *American Men of Science* [1, 8].

3.13   Cornelia Clapp: Zoologist (1849–1934)

I have always had an idea that if you want to do a thing, there is no reason why you shouldn't do it.

One of the women listed in the first edition of *American Men of Science*, Cornelia Clapp (Fig. 3.4) is considered one of the most important zoologists of her day. After attending Mount Holyoke Seminary (before it became a college), Clapp taught mathematics and gymnastics at the school until she was inspired by Lydia Shattuck, Mount Holyoke's science teacher to pursue additional education in the sciences.

Fig. 3.4 Cornelia Clapp.
(Courtesy Smithsonian
Institution)

After spending time at the Anderson School of Natural History on Penikese Island off of Cape Cod, Massachusetts, she returned to Mount Holyoke determined to establish a modern zoology laboratory.

Clapp was actively involved with the Marine Biological Laboratory at Woods Hole, Massachusetts, the successor to the Anderson School, starting at its opening in 1888. She taught there, served as librarian and as a trustee. The research she conducted at Woods Hole was key to the development of the fields of marine biology and embryology. She also continued her education, earning PhDs from Syracuse University and the University of Chicago. She taught at Mount Holyoke all of this time, except during her three-year leave of absence to attend the University of Chicago.

A revered teacher, Clapp became a professor of zoology at Mount Holyoke College in 1904, after establishing the zoology department. A student said, "I came;

I saw; she conquered. Her bounding vitality and thirst for knowledge were contagious. I felt then and have felt ever since that I was never fully alive until I knew her." Clapp retired from Mount Holyoke in 1916, after teaching for 40 years, and continued to summer at Woods Hole.

A Fellow of the Naples Zoological Station in Italy and of the AAAS, Clapp received an honorary doctorate from Mount Holyoke. The laboratory building completed there in 1923 was named for her as was Clapp Road on Woods Hole. Clapp's name was starred in *American Men in Science* [1, 10, 11, 19].

3.14   Sofia Kovalevskaya[15]: Mathematician (1850–1891)

Regarded as one of the greatest mathematical geniuses among women in the last two centuries, Sofia's interest in mathematics seems to have been stimulated by wallpaper at the family estate that consisted of sheets of lithographed lectures on differential and integral calculus. Tutored as a young woman in spite of her father's strong prejudices against educating women, Sofia was determined to go abroad to study as Russian universities were closed to women. She entered into a marriage of convenience with Vladimir Kovalevsky in September 1868 and moved to Heidelberg, Germany. Because she was a woman, she needed to obtain permission from each of her professors allowing her to attend classes.

After three semesters at Heidelberg, she sought out the renowned mathematician Karl Weierstrass at the University of Berlin. Because she was barred from attending classes due to her gender, Weierstrass, who was very impressed with her mathematical abilities, taught her privately twice a week. By the spring of 1874, Sofia had written three doctoral dissertations and was awarded her doctoral degree *summa cum laude* from the University of Göttingen, the first woman to earn a doctorate in mathematics.

Sofia and her husband returned to Russia, had a daughter, and Sofia pursued her other passion, writing. After her husband's suicide, she was offered a position at the University of Stockholm as a privatdozent – a private lecturer who could receive payment from students, but not from the university. She arrived in Sweden in 1883 and by July 1884, she was given a five-year contract as a professor in mathematics. Also in 1884, she became the editor of the new journal *Acta Mathematica*. In 1885, she became the Chair of Mechanics, the first woman to hold a chair at a European university. In 1889, she was given a lifetime appointment to the university. In 1888, Sofia received the Prix Bordin, a prize from the French Academy of Sciences, for substantive contributions to the solution of a long-standing problem in mechanics: the rotation of a solid body about a fixed point.

---

[15] Her name is spelled in many different ways in articles and books referring to her because of its translation from the Russian. It has appeared as Sofia Kovalevsky, Sonia Kovalevsky, Sophia Kovalevsky, Sofya Kovalevsky, Sofia Kovalevskaia, Sonya Kovalevskaia, and Sonia Kovalevskaya.

The research that won her the Prix Bordin is now known as the Kovalevskaya top and her doctoral dissertation on partial differential equations lives on as the Cauchy-Kovalevskaya Theorem [5, 10, 18].

3.15   Hertha Ayrton: Physicist (1854–1923)

Most well-known for the Ayrton Fan, a fan that dispelled poison gases from the trenches in World War I, Hertha Ayrton was the first woman elected to the (British) Institution of Electrical Engineers. Ayrton completed her studies for a baccalaureate at Cambridge University in 1881, but did not receive a degree as Cambridge did not offer degrees to women at that time. She then taught high school mathematics and served as a private tutor.

In 1884, Ayrton invented and patented a line divider. This instrument divides a line into any given number of equal parts. Inspired by this success, she enrolled in Finsburg Technical College in London to study science. Here she worked with W.E. Ayrton, a professor of physics and a pioneer in electrical engineering, whom she married in 1885.

In 1893, Ayrton took over her husband's arc experiments and solved many of the problems with electric arc lamps. She became a recognized authority on the arc. A compilation of her findings was published in book form in 1902. *The Electric Arc* became the standard text in the field.

After moving to the seaside, Ayrton became interested in the action of ocean waves and the formation of sand ripples. She wrote papers on these topics that were of interest to the scientific community. Although refused admittance to the Royal Society in 1902 because of her gender, in 1904, she became the first woman to read her own paper before the group. In 1906, the Society awarded her its Hughes Medal for her work on both the electric arc and sand ripples [10].

3.16   Williamina Paton Stevens Fleming: Astronomer (1857–1911)

The first woman astronomer at Harvard College Observatory to make significant contributions to astronomy, Fleming (Fig. 3.5) was originally hired by Edward Pickering, director of the Harvard College Observatory, to perform clerical and computing tasks. Shortly afterwards, she became a permanent member of the staff and administered the observatory's project of classification of the stars on the basis of the photographed spectra. Fleming was the first woman to receive a corporation appointment at Harvard in 1898 when she was named curator of astronomical photographs.

Fig. 3.5 Williamina Paton
Stevens
Fleming – Wikimedia

Fleming discovered ten novae, or exploding stars, and more than 200 variable stars. She also discovered 59 gaseous nebulae. She developed a new star classification system. Her classification of 10,351 stars in 17 categories for the *Draper Catalogue of Stellar Spectra* was published in 1890. Her system of classification, known as the Pickering-Fleming system, supplanted the previous classification system that had been known as the Pickering system. In 1906, Fleming was elected to the Royal Astronomical Society, one of only four women members. Her name is starred in the first and second editions of *American Men of Science*. Fleming supervised several women who went on to become famous astronomers in their own right including Antonia Maury, Henrietta Leavitt, and Annie Jump Cannon [8, 10].

3.17   Elizabeth Gertrude Knight Britton: Botanist (1858–1934)

The preeminent botanist of the late nineteenth and early twentieth centuries, Elizabeth Gertrude Knight Britton (Fig. 3.6) published more than 300 scientific papers, primarily on mosses – the study of mosses is called bryology. Her first scientific paper on albinism in plants was published in March of 1883 and her first paper on mosses was published later that year. She was the only woman among the 25 charter members of the Botanical Society of American in 1893.

Fig. 3.6 Elizabeth
Gertrude Britton –
New York Botanical
Gardens

Britton contributed to public service as the driving force toward the establishment of the New York Botanical Garden and the preservation of endangered species. She was one of the primary founders of the Wild Flower Preservation Society of America in 1902. She was the principal founder of the Sullivant Moss Society in 1898, renamed the American Bryological Society in 1949. There were 15 species of plants and the moss genus Bryobrittonia named in her honor. In 1935, a double peak in Puerto Rico's Luquillo National Park was given the name of Mount Britton. Her name is starred in the first five editions of *American Men of Science*. Her husband, Nathaniel L. Britton, is also starred. He was originally a geologist but later became an acclaimed botanist in his own right [1, 8, 10].

3.18   Florence Bascom: Geologist (1862–1945)

Florence Bascom (Fig. 3.7) was the first woman to receive a PhD in geology from an American university, the first woman to receive a PhD of any kind from Johns Hopkins University (1893), and the first woman to join the United States Geological Survey (1896). As a professor at Bryn Mawr College, she turned one course in geology into a full major in less than a decade. Her mapping of the crystalline rock formations in Pennsylvania, New Jersey, and Maryland became the basis for many later studies in the area. Her specialty was petrology, the study of how present-day rocks were formed.

Bascom introduced the microscopic study of minerals in the U.S. Bryn Mawr had no facilities for geological research, but Bascom secured rock and mineral specimens and began accepting graduate students from all across the U.S. and Europe. Through the 1930s, a majority of American women geologists had come through her program – of the eleven women fellows of the Geological Society of America by 1937, eight of them had been her students at Bryn Mawr. The first woman fellow of the Geological Society of America (1894), Bascom's name is starred in the first seven editions of *American Men of Science* [8, 10, 20].

Fig. 3.7 Florence
Bascom – Wikimedia

3.19 Zonia Baber: Geographer and Geologist (1862–1956)

Listed in the first edition of *American Men of Science*, Zonia (Mary Arizona) Baber pioneered methods for teaching geography. She was a strong advocate for field work and experimentation such that students became immersed in experiential learning. Baber wrote articles and books explaining her methodology. In addition, to better accommodate student learning in her classes, she invented a desk that had trays and compartments to accommodate learning supplies. In 1896, Baber received patent number 563,638 for her invention (Fig. 3.8).

Baber pursued teaching as a career, following the path that many women of her era took. She gained her education while she was teaching, graduating with a BS from the University of Chicago almost twenty years after she began her teaching career. She headed the geography and geology department in the School of Education at the University of Chicago and also served as principal of the laboratory school at the same institution.

Being passionate about social justice issues as well as education, geography and geology, Baber served as Chairwoman of the Women's International League for Peace and Freedom and as a member of the executive committee of Chicago Branch of the National Association for the Advancement of Colored People (NAACP). She worked for women's suffrage within the continental U.S. as well as in Puerto Rico. She was active in professional associations and served as a founding member of the Chicago Geographic Society. Later, she was president of that organization and received its lifetime award [8, 12, 21].

3.20 Mary Whiton Calkins: Psychologist (1863–1930)

Pioneering psychologist Mary Whiton Calkins completed her work towards a PhD from Harvard University but was denied the degree due to her gender. She established a psychology laboratory at Wellesley College, the first at any of the women's colleges and one of the first in the country. Her areas of interest included the conscious self, dreams, emotions, and memories. She was also interested in philosophy, particularly the area of metaphysics. Over her career, she published 68 papers in psychology and 37 in philosophy and wrote books. In 1905, she became the first woman elected president of the American Psychological Association. Harvard later offered her a PhD from Radcliffe, but she refused. Her name was starred in the first four editions of *American Men in Science* [1, 8].

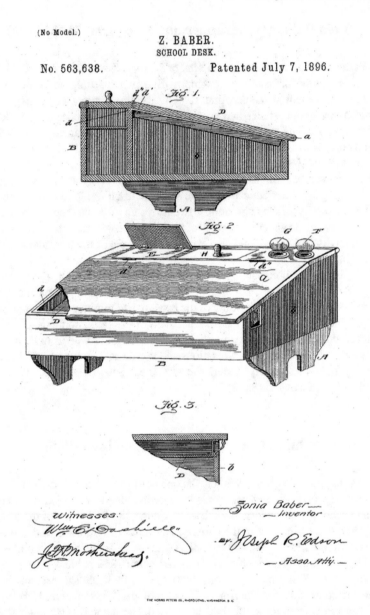

Fig. 3.8 Zonia Baber school desk patent

3.21 Annie Jump Cannon: Astronomer (1863–1941)

American astronomer Annie Jump Cannon (Fig. 3.9) spent her entire career at the
Harvard College Observatory where she developed a cataloging system that arranges
stars by their temperatures and compositions. In her lifetime, she classified the

Fig. 3.9 Annie Jump
Cannon. (Courtesy of
Library of Congress)

spectra of hundreds of thousands of stars for which she received the Henry Draper
Medal in 1931 from the National Academy of Sciences.

The eldest of three children of her father's second marriage, Cannon's mother
taught her the constellations. As a child, Cannon observed the night sky through an
attic window while looking at her astronomical guidebook by candlelight. At
Wellesley, Cannon studied astronomy and graduated with a BS in 1884 before
returning to live at the family home in Dover, Delaware for ten years. Subsequent to
her mother's death, Cannon returned to Wellesley for post-graduate studies and in
1895 she enrolled at Radcliffe College as a special student in astronomy. In 1896,
Cannon became an assistant at the Harvard College Observatory through the support of Edward Pickering, director of the observatory and a leader in the field of
spectroscopy.

There she created a classification system that combined systems developed by
two other female astronomers, Antonia Maury and Williamina P. Fleming. By 1910,
the astronomical community adopted her classification system, known as the
"Harvard system." It is still used today, with minor modifications, and is known as
the Harvard Spectral Classification. She gathered the data for the Henry Draper
Catalogue of stellar spectra which was published by Harvard in ten volumes from
1918 to 1924 and lists the spectral types of over 225,000 stars. Over her lifetime,
Cannon was to classify over 400,000 stars. In 1938, she was appointed William
Cranch Bond Astronomer at Harvard University, one of the first appointments for a

woman by the Harvard Corporation. This occurred very late in her career. The Observatory Visiting Committee had commented on her work during the production of the *Draper Catalogue*:

> At the present time, she is the one person in the world who can do this work quickly and accurately... It is an anomaly that though she is recognized the world over as the greatest living expert in this area of investigation ... yet she holds no official position in the University... It is the unanimous opinion of the Visiting Committee that the University would be honoring itself and doing a simple act of justice to confer upon her an official position which would be a recognition of her scientific attainments.

Cannon received many honorary degrees and awards over her lifetime. She was the first woman to receive the Henry Draper Gold Medal of the National Academy of Sciences (1931), the first woman to receive an honorary doctoral degree from Oxford University, and the first woman elected to an office of the prestigious American Astronomical Society. Her name was starred in the second through sixth editions of *American Men of Science*. She established the Annie J. Cannon Prize of the American Astronomical Society in 1933 to be awarded triennially to a woman who demonstrates distinguished service to astronomy. She was also an advocate for education for women as well as women's suffrage.

Because of her significant accomplishments to the field of astronomy, she was considered for nomination to the National Academy of Sciences. However, no women had yet been elected for membership and her candidacy did not receive adequate support. Posthumously, she was inducted into the National Women's Hall of Fame [8–10, 17].

3.22 Ida Augusta Keller: Plant Physiologist (1866–1932)

Ida Keller was a plant physiologist listed in the first edition of *American Men of Science*. Her doctorate was from the University of Zurich. She taught at Bryn Mawr and then at the Philadelphia School for Girls. Keller's research was in fertilization and in the flora of Philadelphia. She was active in botanical and horticultural societies and published a number of scientific papers as well as a book. Keller was a member of the American Society of Naturalists [1, 8].

3.23 Marie Curie: Physicist (1867–1934)

Perhaps the most famous of historical women scientists, and to many Americans the only woman scientist's name that they recognize, Marie Sklodowska Curie (Fig. 3.10) was the first woman to receive a Nobel Prize. She was also the first person to become a Nobel Laureate twice, the first person to win in multiple sciences and the only woman to win in two fields. She received the Nobel Prize in Physics

Fig. 3.10 Marie Curie.
(Courtesy Library of
Congress)

1903 jointly with her husband Pierre and Antoine-Henri Becquerel, both physicists. She received the Nobel Prize in Chemistry in 1911.

Marie and Pierre Curie introduced the concept of radioactivity to the world. She also isolated two new radioactive elements, polonium and radium. However, she was never accepted to membership in her country's most prestigious scientific society, Académie des Sciences (one of the five academies of the Institut de France), because "women cannot be part of the Institut de France."

Marie Sklodowska was born in Warsaw, Poland. She began work as a governess at age seventeen to pay for her older sister's medical education. When her sister had finished medical school, Marie went to Paris to join her. In 1891, at the age of 24, she enrolled at the Sorbonne, being one of the few women in attendance at the university. She received a degree in physics in 1893, graduating first in her class, and in 1894, received her master's degree, this time second in her class. In 1894, she met Pierre Curie, a physicist, and the two were married in 1895.

By the time she had completed the work for her doctorate in 1903 (*summa cum laude*), the first woman in France to earn a doctorate, Marie and Pierre had discovered radioactivity and had identified and isolated polonium and radium. After Pierre was killed in a tragic accident in 1906, Marie was invited to take over his teaching position at the University of Paris. When she accepted, she became the first woman to receive a post in higher education in France, although she was not named a full professor for two more years.

Curie volunteered at the National Aid Society during World War I and brought her radiology technology to the war front. After determining that soldiers would have the best chance of recovering from their wounds if operated on quickly, she developed mobile x-ray facilities that could be brought near the front lines. These units came to be known as petite Curies ("little Curies"). In addition, she became the director of the Red Cross Radiology Service. Curie set up the first military radiology center for France. Her daughter Irene served as one of her first assistants. It is estimated that more than one million soldiers were treated with her x-ray units.

By the 1920s, Curie was an international figure. She established the Curie Foundation in 1920 to accept private donations for research. One of her significant and almost unknown legacies is that her daughter Irene Joliot-Curie followed in her footsteps, receiving a doctorate in physics and becoming a Nobel Laureate herself [3, 9, 10, 12, 22].

3.24   Caroline Furness: Astronomer (1869–1936)

Pioneering astronomer Caroline Furness was hired by Mary Whitney in 1894 as her assistant at the Vassar College Observatory. Furness would later succeed her as its director and professor of astronomy holding the Maria Mitchell Professor of Astronomy. Furness and Whitney worked together in observations of comets and minor planets and later, on variable stars. Furness emphasized the use of photography to assist in astronomical research. Elected a fellow of the Royal Astronomical Society, Furness was also a member of the American Association for the Advancement of Science. Her name is included in the first edition of *American Men of Science* [1, 8].

References

1. M.W. Rossiter, *Women Scientists in America: Struggles and Strategies to 1940* (The Johns Hopkins University Press, Baltimore, 1992)
2. B. Harris, *Beyond her Sphere: Women in the Professions in American History* (Greenwood Press, Westport, 1978)
3. S.A. Ambrose, K.L. Dunkle, B.B. Lazarus, I. Nair, D.A. Harkus, *Journeys of Women in Science and Engineering: No Universal Constants* (Temple University Press, Philadelphia, 1997)
4. "Elmira College: About the College". www.Elmira.edu/aboutco/GLANCE.HTML. Accessed 31 Sept 2001
5. H.J. Mozans, *Woman in Science: With an Introductory Chapter on Woman's Long Struggle for Things of the Mind* (University of Notre Dame Press, Notre Dame, 1991)

6. "A Brief History of Barnard". www.barnard.edu/about/history.htm. Accessed 11 Mar 2002
7. "Brown University History: Part Three, Building a University". www.brown.edu/webmaster/about/history/part3.shtml. Accessed 11 Mar 2002
8. M.J. Bailey, *American Women in Science: A Biographical Dictionary* (ABC-CLIO, Denver, 1994)
9. B. F. Shearer, B. S. Shearer (eds.), *Notable Women in the Physical Sciences* (Greenwood Press, Westport, 1997)
10. P. Proffitt (ed.), *Notable Women Scientists* (The Gale Group, Detroit, 1999)
11. American Men and Women of Science. https://en.wikipedia.org/wiki/American_Men_and_Women_of_Science. Accessed 9 May 2020
12. M.B. Ogilvie, *Women in Science: Antiquity through the Nineteenth Century, a Biographical Dictionary with Annotated Bibliography* (MIT Press, Cambridge, 1993)
13. G. Kass-Simon, P. Farnes (eds.), *Women of Science: Righting the Record* (Indiana University Press, Bloomington, 1990)
14. Merriam-Webster, Inc, *Webster's Dictionary of American Women* (Smithmark Publishers, New York, 1996)
15. P.J. Read, B.L. Witlieb, *The Book of Women's Firsts* (Random House, New York, 1992)
16. "Ellen Richards". www.greatwomen.org/rchrdse.htm. Accessed 26 May 1999
17. E. Yost, *American Women of Science* (J.B. Lippincott Company, Philadelphia, 1955)
18. C. Morrow, T. Perl (eds.), *Notable Women in Mathematics: A Biographical Dictionary* (Greenwood Press, Westport, 1998)
19. B. F. Shearer, B. S. Shearer (eds.), *Notable Women in the Life Sciences* (Greenwood Press, Westport, 1996)
20. L. Barber, *Four Lives in Science: Women's Education in the Nineteenth Century* (Schocken Books, New York, 1984)
21. Z. Baber. https://en.wikipedia.org/wiki/Zonia_Baber. Accessed 11 May 2020
22. M. Curie. https://en.wikipedia.org/wiki/Marie_Curie. Accessed 9 May 2020

Chapter 4
War Brings Opportunities

4.1 World War I

During World War I, women were encouraged to participate in the work force and support the war effort (Fig. 4.1). Some women took temporary jobs during the war, filled in for men serving in the military, or performed volunteer work either at home or abroad. Women took jobs in blast furnaces, in the manufacture of steel plate, high explosives, armaments, machine tools, agricultural implements, electrical apparatus, automobile, airplane, and railway parts. Women worked in brass and copper smelting and refining; in textile mills producing uniforms for the armed services. Women could be found in foundries, in oil refining, in the production of chemicals, fertilizers, and leather goods. And women could be found in the transport services and other occupations [1, 2].

The war opened up jobs in industry for women scientists for practically the first time and expanded women scientists' employment opportunities in government. It also created activity and turmoil on college campuses [1].

The U.S. government had not made much preparation for war and thus the spring of 1917 was chaotic. No priorities had been established for the optimal use of manpower, so industries and agencies competed heatedly for the limited men available. Intense demand was experienced for industrial chemists, to increase the American production of war materiel, including gases and, as well as to make up for the loss of chemical imports from Germany on which the U.S. had become significantly dependent. Male chemists did not exist in significant numbers to meet this demand so women chemists who had previously been actively discouraged or excluded from the field were now encouraged to enter the industry. Many women chemists were hired by industry during 1917 and 1918 and their efforts were praised [1]:

> *They have added tone to our laboratory by their pleasing personalities. They have proved beyond a doubt that they can do and will do at any hour of the day or night, careful, conscientious, reliable, chemical work. They have passed the crucial test of service. They have been weighted in the chemical balance and not found wanting.*

J. S. Tietjen, *Scientific Women*, Women in Engineering and Science,
https://doi.org/10.1007/978-3-030-51445-7_4

Fig. 4.1 Women Needed for the World War I Effort. (Courtesy of Library of Congress)

Some women were able to obtain government scientific jobs, although most of the jobs available appear to have been filled through the old-boys network and many were within the ranks of the military from which women were precluded. The National Bureau of Standards hired physicist Louise McDowell to work on radar in 1918. Physicist Frances Wick worked on airplane radios and gun sights for the Army's Signal Corps for a while in 1918. The Office of the Chief of Staff hired Louise Stevens Bryant to write statistical reports on food supplies available for both the army and the Allies[1] [1, 3].

Interestingly, the area in which women scientists appear to have made their most significant contribution was in that area of science derided as "women's work" –

---

[1] Louise McDowell, whose PhD was from Cornell University, spent her academic career at Wellesley. Frances Wick and Louise McDowell met at Cornell from which Wick also got her PhD. Wick's academic career was spent at Vassar College. Louise Stevens Bryant was a physician.

home economics. The U.S. Food Administration was established in 1917 to supplement efforts of the Department of Agriculture because of predictions of serious food shortages in Europe in the year ahead. The charge for this agency was to figure out how to increase food production and decrease food consumption through means short of rationing. Although a woman was not selected to head the agency, several women held high-ranking positions within the organization and hired women as staff and as consultants, Mary Engle Pennington and Mary Swartz Rose among them, to assist in the task. They were successful: American consumption of food dropped about 10% in 1918 freeing food for export to the Allies in Europe [1].

Women scientists on college campuses offered war-related courses and training in "women's work." Several colleges offered courses on nutrition and food preservation, industrial chemistry, bacteriology, mapmaking, and wireless telegraphy. Mount Holyoke trained women inspectors about health conditions in industrial plants. Women students were urged to till "war gardens" and to serve as agricultural laborers "farmerettes" during the summers of 1917 and 1918. Geologist Ida Helen Ogilvie's family farm, Airlie, became the first Women's Land Army unit in the U.S. and she toured colleges recruiting the farmerettes (Fig. 4.2) [1, 3].

One category of women scientists especially benefited from the influenza pandemic in 1918. Public health personnel, particularly bacteriologists, were required as the country was experiencing a fall-scale public health crisis. Science majors were recruited from the eastern women's colleges for summer internships with the promise that such would lead to professional positions in laboratories and inspectorships [1].

At least one source of career information in 1919 was not overly encouraging for women considering careers as bacteriologists, however [4].

Not the least among the remarkable array of professions open to women is that of bacteriologist. This profession will appeal to the girl who has a systematic turn of mind, an ability to learn new things quickly, accuracy and conscientious attention to duty, together with a love of discovering new truths and an absolute lack of sentimentality.

These sound like a formidable array of almost impossible qualifications, but in reality one leads directly to the other. If a girl likes laboratory work and desires to enter that field more than any other, she ought to be successful. Some girls begin work in a laboratory and after two or three months find that instead of absorbing their attention the work only weakens and sickens them...

There isn't any definite plan of promotion or increase in salary. Promotion comes only through death of a person holding a high position or through a resignation... Unless a girl desires to work in a laboratory more than anything else in the world, she ought not to attempt the work... The recompense is small in comparison to the labor involved. If you don't love the work, it soon becomes drudgery. You have to forget self. Most girls faint the first time they assist at a dissection, but they soon get over the feeling.

A very significant event in the women's rights movement occurred just after World War I – women achieved the right to vote. Many historians believe that the vote for women's suffrage, ratification of which was finally achieved in 1920, was in gratitude for women's efforts during the war. However, in spite of that "gratitude," most women were not allowed to keep the jobs they had filled during the war. In fact, they were commended for yielding their jobs to the returning GIs: "The

Fig. 4.2 Farmerette. (Courtesy Library of Congress)

women chemists of the Illinois Steel Company not only made good as chemists but showed their fine spirit by resigning in order to make places for the men returning from war work" [1].

It was time for women scientists to celebrate and capitalize on the theme of women and science and an opportunity arose – Marie Curie, Nobel Laureate, was invited to come to the U.S. on tour! In 1921, the "First Lady of Science" (Fig. 4.3) toured the U.S. for three weeks, visiting the White House and President Harding, receiving more than twenty honorary degrees, and being provided with a gram of radium worth over $100,000 with which to continue her research. However, U.S. women scientists benefited marginally from these efforts and would see very little advancement in the opportunities they were afforded over the next two decades [1].

Fig. 4.3 Marie Curie at the White House with President Harding. (Courtesy of Library of Congress)

By 1927, women who were interested in scientific careers were being given vocational advice that tended to track them into women's work and didn't offer much hope for advancement or pay in non-traditional fields [5].

Women trained for scientific work are profiting by the fact that many of the more recent developments in scientific research and in its applications to living, concern interests traditionally within women's sphere.

Anthropology: This is an open field for women but not an especially promising one, and the number of women working in the field is small.

Archaeology: Women, in small numbers, have, for a long time, been active and successful in archeology, a considerable number having won fellowships for foreign study and having helped in important discoveries.

Astronomy: A few women have attained distinction in astronomy. Outside of teaching, women are employed chiefly as research computers and as assistants in observatories, and in scientific libraries...It seems wise for only a very small proportion of women to prepare to enter the field.

Biological science: A biological field as yet but little exploited by women, but one for which their natural tastes and talents might be expected to equip them, is landscape architecture.

Botany: Inasmuch as women are credited with some especial aptitude in relation to plants and flowers, some advantage obtains for well equipped woman botanists, but their work still lies much more with plants and flowers, than with larger botanical developments, or with the more significant application of these to human problems... few have won the highest professional recognition, even to the extent of holding membership in the Botanical Society of America...

Zoology: Zoological expeditions offer the largest interest afforded by zoological occupations but are somewhat rarely opened to women. This condition may pass, however, as the sturdy physical type develops among women.

Chemistry: Women are employed to some extent in all these connections [teaching, research, and applied chemistry], *and although their equipment far too often qualifies them only for routine tasks, opportunity for the higher positions is often lacking or difficult to secure...Men hold most of the more important teaching positions in the highly accredited women's colleges, as well as in many others.*

Geology: Geological field work is usually called unsuitable for women, but some have disproved this for themselves by their physical endurance and power of adaptation. This proof seems likely to be given increasingly in the future, as women develop more vigorous physique.

Mathematics: In colleges where the faculty includes both sexes, the better opportunities tend, as usual, toward the male teachers, ...

Physics: The number of highly equipped women physicists or of those in responsible positions, has grown more slowly than is true in the fields of biology, chemistry or psychology... In these again [physical research opportunities], *women are seldom engaged in the more advanced types of work...Women have made a substantial beginning, but hardly more, in occupations using applied physics... Women physicists as a rule are in minor positions in this field* [teaching physics in higher education].

Psychology: A considerable number of the psychologists connected with child welfare institutes are women...Women have an encouraging opportunity to share in such work [psychology applied to understanding physical and mental illness], *where it is concerned with women and children... Women have special aptitude for certain phases of such work* [juvenile research], *and an appreciable number of them are already engaged in it. Opportunities for women in remunerative psychological research as a full time occupation, are not numerous... For many of the applications of psychology to education, women have a certain sex advantage in their special understanding of children, and in their consequent ability to work well with them... Another opening to which women are usually eligible, is found in health guidance clinics being developed in the larger cities.*

Laboratory Technicians: So large a proportion of women associated in any way with scientific work are performing the duties of laboratory technicians or clinicians... in spite of the fact that relatively few laboratory technicians are scientists. An overwhelming proportion of them are mechanicians only.

Women scientific workers, not yet having achieved many of the higher positions – whether in teaching, research business, or industry – are not, as a rule, among the more highly paid ... although the scale is lower... for women than for men.

A few women have made striking advances in science, but the uncompromising discipline and concentration required make success of the first rank for them in science or research much more self-abnegating on the human side than for men, because of the home interests which they must conserve for themselves and which are usually safeguarded for men by their wives.

4.2   Great Depression

Men and women alike were forced to cope when the stock market crashed in 1929, and the ensuing Great Depression arrived. Many women had to sacrifice personal ambition to accept a life of economic inactivity [6]. Thus many women either voluntarily or involuntarily ceased their careers for a period of time after 1929. Jobs in general were scarce and what jobs were available were unlikely to be filled by women.

Women providing career advice to young girls in 1929 presented diverse views of their careers based on their experiences [7].

Agriculture: Agriculture does not, upon casual consideration, recommend itself as a vocation for women.

Home Economics: Competition is with women, hence the opportunity for advancement is greater than in a subject where men tend to hold all the higher positions.

Science: Science as a career for women is practically the same as it is for men. Although the supervision of scientific research is still largely in the hands of men, many of the actual experiments are done by women... Many of the women in these branches [biological and chemical sciences] *are employed merely as technicians.. .A few who are better qualified... aspire to the higher positions... The small financial return, because salaries are small when compared with those of other fields, is more than compensated for by the real joy of the work and the glory subsequent to a definite contribution... When competing with men for a position or promotion, especially in the industrial world, the preference will be given to men. In order to overcome the traditional prejudice or favoritism, for it does exist, the writer has always urged that women acquire a superior education and training...*

The scientific labor force, however, grew tremendously between 1920 and 1940. More and more female and male scientific doctorates were being graduated and the number of positions was increasing both at leadership and subordinate levels as scientific teaching, research, and application attracted new funding from universities, industry, foundations, and all levels of government. Female scientific doctorates increased from about 50 per year nationwide in 1920 to about 165 per year nationwide by the late 1930s. Because not only women with doctorates were characterized as scientists, it is useful to look at data from the editions of *American Men of Science* between 1921 and 1938. In 1921, 4.7% of the more than 9000 persons listed were women (450 women) and by 1938, that percentage had increased to 7.0% of more than 27,000 persons (1912 women). Most of these women scientists were employed in higher education and had suffered the brunt of layoffs and pay cuts when universities needed to reduce costs during the depression. Many of the women in the chemical industry also found themselves without employment in the 1930s or were forced to take jobs considered "women's work" to remain employed [1].

At the end of the decade, women's career advice still sounded similar to that advocated in earlier decades. After a pretty harsh introduction, careers were described in terms of salary, work opportunities, and chances for promotion [8]:

> She who takes science for her mistress chooses truth... The history of science is filled with the names of martyrs who died rather than recant their belief in the truth they had learned from research... Modern science has its martyrs no less than those of the Middle Ages. To-day men give their lives in the battle with disease; men sacrifice position and wealth to pursue the transient gleam of scientific discovery.
>
> Can a woman carry both burdens [work and family]? The answer is as final as the physical law that two things cannot occupy the same space at the same time. She cannot give first attention to both... She must recognize these limitations and plan her life accordingly, for she has no right to take on responsibility which she cannot fulfill.
>
> Bacteriology and Medical Research: The opportunities for women are particularly good in the civil-service branches of this field and in hospital work, but they rarely attain the choice positions on medical faculties and in private foundations... Until recently women in government service were admitted to the study of plant and human diseases but barred from animal infections, a regulation worthy of note simply as an example of the fact that a sense of humor is a satisfying possession in woman's struggle for a place in the industries.
>
> Chemistry: The woman in chemistry may find some special interest in the branches of food chemistry, textiles, and dyes, and similar work more closely related to the household arts. Her opportunity to advance in these branches is greater than in the general industrial field, should they make a genuine appeal to her, as there is less prejudice against women here. Teaching Chemistry:... there is little hope of her advancing to the chairmanship of a department or conducting real research work...There is perhaps more opposition to women in chemistry than in any other science. [Men] feel a constraint in speech and dress with women in the laboratory which interferes with their work....
>
> Physics: Only about one-fifth as many women find a life interest in physics as in chemistry. This is partly because there are fewer openings in the profession and partly because the mathematical character of the work appeals less to women. . . the higher paid positions of the profession are not, as yet, open to them.

4.3 World War II

World War II brought about tremendous economic and social changes in the U.S. The vestiges of the Great Depression were almost eliminated with the sudden high employment. Women were needed to support the war effort (Fig. 4.4). The increased need for governmental services that accompanies war efforts led to increases in the size and responsibility of the federal government (including more jobs). Government's interest and support of science also tends to increase during wartime and World War II was no exception. Women scientists had increased opportunities for employment and visibility during World War II. For example,

Fig. 4.4 Women Needed
for the WWII Effort.
(Courtesy of Library of
Congress)

The more **WOMEN** at work
the sooner we **WIN!**

WOMEN ARE NEEDED ALSO AS:

FARM WORKERS	WAITRESSES	TIMEKEEPERS	LAUNDRESSES
TYPISTS	BUS DRIVERS	ELEVATOR OPERATORS	TEACHERS
SALESPEOPLE	TAXI DRIVERS	MESSENGERS	CONDUCTORS

— and in hundreds of other war jobs!

SEE YOUR LOCAL U.S. EMPLOYMENT SERVICE

corporations including Monsanto, Du Pont, and Standard Oil hired women chemists for the first time. Unfortunately, many of the opportunities were temporary – placeholders until the boys came home from war [9, 10].

The movement of women into the work force during World War II is often personified by famed metalworker and poster woman "Rosie the Riveter (Fig. 4.5)." Rosie is portrayed as a powerful woman, pictured in a headscarf with sleeves rolled up stating "We Can Do It!" [6, 9] And women scientists were needed.

The Office of War Information (OWI) and the War Manpower Commission began issuing literature and propaganda glorifying women as scientists and engineers to bolster the war effort. In 1942, a film, *Women in Defense*, narrated by award-winning actress Katharine Hepburn reading a script written by First Lady Eleanor Roosevelt, urged women to go to work in government or on scientific projects. In 1942, the movie *Madame Curie*, starring award-winning actress Greer Garson, was released, further glorifying women's contributions in science. New training programs for scientists and engineers were established and by 1943, women

Fig. 4.5 Rosie the Riveter.
(Courtesy National
Archives and Records
Administration)

and to some extent, blacks, were being specially recruited and trained for jobs in industry. Bright high school students were sought out and urged to major in science in college through the Westinghouse National Science Talent Search (established 1943) and the Bausch and Lomb Science Talent Search (established 1944). Both programs had women among their early winners and the Westinghouse program required that there be the same percentage of women winners as there were women entrants. Books and articles were released during the 1942 through 1945 period urging women to pursue careers in science and engineering. Edna Yost's book *American Women of Science* (1943), which lauded women's past and current contributions to science, painted a bright picture for women in these fields[2] [9].

Women were encouraged to enter the work force because the men were gone and not just in the areas where women were already in the work force; those invisible

---

[2] However, even Yost expressed surprise in her preface to *American Women of Science* that women had already done so much: *But I had not the slightest idea that American women of science had achieved even a fractional part of what they have actually accomplished. I was completely unprepared for what a little specialized research began to uncover. So, it seems, was everybody else. When we signed the contract, both my publisher and I knew there was material available for a pretty good book but we had no idea the achievements recorded would be of the caliber they actually are.*

jobs discounted as "women's work"[3] [11]. People were needed everywhere that the war effort needed support. With the men gone, women and blacks, the two reserve labor forces of the country were needed to fill these positions. But, there was a problem: not enough women or blacks with the relevant training were ready to fill the vacant jobs [9, 12].

To fill the personnel gap, the U.S. government began running crash courses in science, engineering, and management for both men and women after the war began [12]. However, some programs and associated literature were still directing women into lower level positions or otherwise not encouraging women's ambitions for career success in scientific fields. A 1943 booklet, *Civilian War Service Opportunities for College and University Students*, described about one hundred jobs and the prospects for women being employed in such jobs. A positive outlook was indicated for statisticians: "Women with statistical training and appropriate supporting experience are in great demand." Physicists were not provided with as much encouragement: "Women should be encouraged to enter the field of physics. There are a large number of openings for women, particularly in the lower grades." And those women aspiring to careers in the agricultural fields were pretty much discouraged: "Women have not held many positions [junior soil conservationist] of this type." and "These positions [high-level agricultural specialist] will be filled primarily by men" [9].

4.3.1 Faculty

Women also were encouraged to become faculty members at colleges and universities, temporarily replacing men who had been called into government service or into the military. In fact, the number of women scientists, including mathematicians and chemists who were not working on other war projects, moving into the ranks of college and university faculty was highly significant and probably constituted women's greatest contributions to the war effort. In 1942, 2412 women scientists were on science faculties. By 1946, the number had increased to 7746 – an over 200 percent increase in 4 years! Women had the opportunity to teach at larger institutions and all-male colleges, and some even had the opportunity to become department chairs. However, many large universities placed a moratorium on tenure decisions until after the war – so these patriotic women could not expect the jobs would be other than temporary [9, 11].

---

[3] One ready source of female labor was from the curtailment of female-intensive industries unnecessary to the war effort, the largest being silk hosiery manufacturing. The Women's Bureau of the U.S. Department of Labor developed a policy regarding the order in which women should be employed: unemployed women workers displaced by manufacturing changes, girls from high school and college, then homemakers (presumably meant to include older women), and finally mothers of small children as a last resort. By July 1943, there were 17.7 million women in the civilian labor force; there had been 13.9 million in June 1940.

Discrimination against women and, specifically women faculty, did continue after the war ended. The 1947 Steelman Report to President Truman on immediate and future scientific needs of the nation surveyed science faculty across the country. Sample comments revealed the high hurdles that women still had to overcome in order to be employed, never mind achieving acceptance: (1) 40.9% of the mathematics chairs responded that the gender of the professor did not matter but 22.7% were definitely opposed to women faculty and (2) geologists claimed they were generally indifferent, but 18% stated that women would be severely handicapped in this career choice because of the requirement for field work [9].

4.4   After World War II

When the war ended, millions of veterans, primarily men, came home expecting to be gainfully employed. Many of the women who had been employed found themselves no longer welcome in the work force. In fact, these women were now expected to go home and raise babies or "go back to the kitchen" and not steal work from returning GIs. The special educational programs were shut down and educational opportunities for women were again limited. By January 1946, four million women had left the labor force [6, 11].

By the time World War II ended, however, women were not quite as eager to accept the inevitable as they had when World War I ended. They began generating statistics and scientific data to prove they were just as capable as men and were being treated unfairly. Still, even for those who managed to hold on to their positions, they were unable to advance as men did. Women were still channeled into less challenging work, kept in lower ranks, and paid lower salaries by employers who minimized their contributions [9].

A very significant impact of the war's end was the GI Bill that provided funds for veterans' education. Schools that had been underutilized during the war were suddenly faced with a deluge of male students – it is estimated that 7.8 million veterans chose to take advantage of the GI Bill's educational provisions. Female enrollment at some coeducational schools had reached 50 or percent higher during the war, but as the demand for slots increased, many institutions reintroduced, or introduced, maximum quotas on female enrollment. Some colleges refused to accept applications from out-of-state women. Women were told that space was not available and that they would have to attend other institutions. Through the 1950s, the percentage of women dropped to just 25–35% of the total college undergraduate enrollment and the remasculinization of science of the 1950s and 1960s was well underway[4] [9].

---

[4]The Serviceman's Readjustment Act of 1944 (commonly known as the GI Bill) included educational provisions offering honorably discharged veterans whose schooling, college, or professional education had been interrupted by military service up to five years of full tuition plus a subsistence allowance at the college of their choice (a very important provision, this last). The nation's 400,

Discrimination against women was widespread; women were systematically pushed out of science and engineering at the undergraduate level, at the graduate level, and as faculty. Men even took over women's dormitories and many universities added temporary housing to accommodate the returning male GIs. Male veterans replaced female staff and faculty as well as female students. There were no discrimination laws in place (these would not be enacted until the 1960s or later) to preclude any of this behavior. The nation was also extremely grateful to its veterans and felt there was, in many cases, a duty to make their adjustment to civilian life as straightforward as possible. Objections or protests were few and far between and not effective. In addition, society as a whole was still quite ambivalent about the proper role for women, especially achieving women or scientific women who were certainly atypical and threatened presumed male and female spheres[5] [9, 11].

The sentiments expressed by the Women's Bureau of the Department of Labor in The *Outlook for Women in Science* (1949) demonstrated this ambivalence [13]:

> Is our Nation finding and developing all its potentially great scientists?. .. The Women's Bureau believes that, although women will probably never equal the number of men in scientific fields, they can and will play an increasingly greater role, both quantitatively and qualitatively, in the sciences... The Bureau presents this factual report as a basis for its belief and its initial contribution both to the increasing number of women who want to train for scientific work and to those who are concerned with the present and potential use of a relatively unmined source of scientific talent.

The 1946–1947 data presented in the report (Table 4.1) for men and women in the scientific fields reported a high of 5400 women in chemistry as compared to 71,600 men and an estimate of 10 women in animal husbandry as compared to 2380 men. The fields with the highest penetration of women were bacteriology (25%), general botany (22.7%), and mathematics (exclusive of statistics) (20.1%) [13].

Women constituted about 3% of the total individuals in all of the scientific fields. It would take the launch of Sputnik and federal legislation before the attitudes of society would change and the numbers and percentage of women in the sciences would increase significantly [13].

---

000 women veterans were also eligible for these benefits, but were often overlooked. In the post-World War II period, quotas and enrollment restrictions on women at colleges and universities enabled male veterans to displace female applicants. In a short period of time, men had also displaced female staff and faculty. For example, during the wartime, women had constituted 50% of enrollment at Cornell University. Through quotas, that number was pared to 20% of the 1946 enrollment.

[5] One dean of women at Indiana University, without warning or advance consultation, was simply ousted from her office, demoted, and given a much lesser title by the new dean, a former military officer. Nepotism rules that had been relaxed during the War were reinstated leading to the dismissal of employed spouses (almost always the wives) of university employees. Faculty women who married immediately lost their jobs.

Table 4.1 Estimated Number and percent distribution of men and women in principal scientific fields in the U.S., 1946–1947

Scientific field	Number			Percent			Percent women are of total
	Total	Men	Women	Total	Men	Women	
All fields listed below	477,890	465,130	12,760	100.0	100.0	100.0	2.7
Architecture	15,000	14,700	300	3.14	3.16	2.35	2.0
Astronomy	600	500	100	0.12	0.11	0.78	16.6
Bacteriology	4000	3000	1000	0.84	0.65	7.84	25.0
Biology, general (exclusive of bacteriology, botany, zoology)	3200	2600	600	0.67	0.56	4.70	18.8
Botanical science:							
General botany	1100	860	250	0.23	0.18	1.96	22.7
Plant physiology and pathology	1050	980	70	0.22	0.21	0.55	6.7
Agricultural plant sciences including forestry	7850	7820	30	1.64	1.68	0.23	0.4
Chemistry	77,000	71,600	5400	16.11	15.39	42.32	7.0
Engineering	317,000	316,050	950	66.33	67.95	7.45	0.3
Geography	800	660	140	0.17	0.14	1.10	17.5
Geology	11,000	10,670	330	2.30	2.29	2.59	3.0
Mathematics (exclusive of statistics)	10,200	8150	2050	2.13	1.75	16.07	20.1
Meteorology	2800	2770	30	0.59	0.60	0.23	1.1
Physics	18,450	17,550	900	3.86	3.77	7.05	4.9
Zoological science:							
General zoology	3900	3470	430	0.82	0.75	3.37	11.0
Physiology	900	790	110	0.19	0.17	0.86	12.2
Pathology	650	590	60	0.14	0.13	0.47	9.2
Animal husbandry	2390	2380	10	0.50	0.51	0.08	0.4

* * *

KEY WOMEN OF THIS PERIOD

* * *

4.5   Florence Sabin: Anatomist (1871–1953)

Florence Rena Sabin (Fig. 4.6), an anatomist and the first woman to be elected to membership in the National Academy of Sciences, is regarded as the outstanding woman scientist in the medical field in the first half of the twentieth century. Her other many firsts included first woman faculty member at Johns Hopkins University (1902), first woman to become a full professor in the medical school at Johns

Fig. 4.6 Florence Sabin.
(Courtesy Library of
Congress)

Hopkins University (1917), first woman elected a member of the Rockefeller
Institute, and the first woman president of the American Association of Anatomists
(1924–1926). Her studies of the central nervous system of newborn infants, the
origin of the lymphatic system, and the immune system's response to infections –
particularly the bacterium that causes tuberculosis – were important in the annals of
science.

Sabin was born in Colorado and attended Smith College with her older sister.
She was particularly interested in mathematics and science and earned a BS in 1893.
A zoology course in her junior year ignited a passion for biology. After completing
medical school, she was accepted as an intern at Johns Hopkins, a rare occurrence
for a woman, and decided that she preferred research and teaching to practicing
medicine.

When Sabin retired, she moved to Denver where she became very active in public health issues. Her name was listed in the first edition and starred in the first
through seventh editions of *American Men of Science*. She received many honorary
doctorates and other awards recognizing her significant accomplishments. Sabin
has been inducted into the National Women's Hall of Fame. She represents Colorado
in National Statuary Hall in the U.S. Capitol in Washington, D.C [14–16].

4.6 Margaret Washburn: Psychologist (1871–1939)

Margaret Washburn, the second woman to be elected to the National Academy of Sciences, was a pioneer in the field of psychology. Her early studies of animal behavior, vision, and speech introduced this new scientific discipline to a generation of students. Her most important publication *The Animal Mind* (1908) analyzed the large literature of animal psychology. Her motor theory of consciousness, that all thoughts and perceptions produce some form of motor reaction, was explained in her book *Movement and Mental Imagery*.

Washburn earned her BA in 1891 in biology, chemistry, and philosophy from Vassar College. She became interested in experimental psychology and applied to study at Columbia University. The trustees refused to admit a woman as a regular graduate student, however, and Washburn was allowed only as a "hearer" or auditor. She subsequently went to Cornell University where women were not only admitted, but also provided with scholarships. She was awarded her doctorate in 1894.

Washburn taught at Wells College, Cornell University, and the University of Cincinnati – where she served as head of the psychology department – before returning to Vassar College where she remained until her retirement in 1937. Her name was listed in the first edition of *American Men of Science* and starred in the first six editions [15, 16].

4.7 Harriet Boyd Hawes: Archeologist (1871–1945)

Growing up in an all-male household after her mother died when she was an infant, Harriet Boyd Hawes expected the opportunities that boys had. She graduated from Smith College and eventually went to the American School of Classical Studies (ASCS) in Athens, Greece for graduate work. During her excavation work on Crete, she discovered tomb sites from the Iron Age, providing her with the subject for her master's thesis which she submitted to Smith. After receiving her master's degree, she taught at Smith for six years, returning several times to Crete to do field work.

The first woman archeologist to head an excavation, she discovered Gournia, one of Crete's "ninety cities" of Homer's *Odyssey*. There, using the money from a fellowship she won and because she wasn't allowed to participate in excavations sponsored by the ASCS due to her gender, she oversaw the excavating work during the years 1901–1904. She published papers and gave a national lecture tour on the Minoan Early Bronze Age town site that she uncovered. After years of raising children and being a social activist, Hawes taught at Wellesley College from 1920 to 1936 [3, 15].

4.8   Mary Engle Pennington: Chemist (1872–1952)

Mary Engle Pennington developed standards of milk and dairy inspection that were adopted by health boards throughout the country. Her methods of preventing spoilage of eggs, milk, poultry, and fish were adopted by the food warehousing, packaging, transportation, and distribution industries. She has six patents associated with refrigeration and spoilage prevention methods (Fig. 4.7). The standards she established for refrigerated railroad cars, based on the understanding she gained by riding freight trains, remained in effect for many years and gained her worldwide recognition as a perishable food expert.

Pennington completed the coursework for a bachelor's degree in chemistry, biology, and hygiene at the University of Pennsylvania, but at that time (1892), the University did not grant bachelor's degrees to women. Instead, she received a Certificate of Proficiency in biology. She continued her studies and, in 1895, received a PhD in chemistry from the University of Pennsylvania.

Her work in refrigeration led to her appointment as head of the Department of Agriculture's food research laboratory. As she used the name "M.E. Pennington," not everyone was aware that she was a woman. In 1916, when she had been chief of the Food Research Laboratory for a decade, a railroad vice-president on whom she called, instructed his secretary "to get rid of the woman," because he had "an appointment with Dr. Pennington, the government expert."

Pennington received the Garvan Medal from the American Chemical Society in 1940 and was the first woman elected to the American Poultry Historical Society's Hall of Fame (1947). In that same year, she was elected a fellow of the American Society of Refrigerating Engineers and a fellow of the American Association for the Advancement of Science. Posthumously, she has been inducted into both the National Women's Hall of Fame and the National Inventors Hall of Fame [17–20].

4.9   Mary Swartz Rose: Chemist and Nutritionist (1874–1941)

After receiving her PhD in chemistry from Yale University, Mary Swartz Rose taught at Columbia University's Teachers College. She would spend the rest of her academic career there in the field of nutrition. It was also there that she established the department of nutrition. Her books included *Laboratory Handbook for Dietetics*, *Feeding the Family*, and *Everyday Foods in War Time*. She published more than forty papers, most of them on the practical aspects of nutrition. She was very interested in nutrition in the public schools and published a book on the subject titled *Teaching Nutrition to Boys and Girls*. During World War I, Rose served as the director of the Bureau of Conservation of the Federal Food Board and the New York State Food Commission.

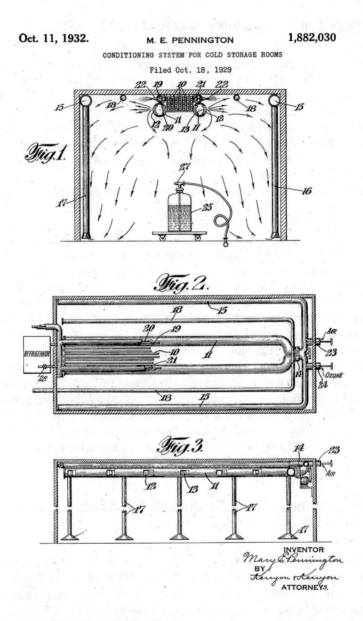

Fig. 4.7 Patent of Mary Engle Pennington

Rose was instrumental in nurturing the science of nutrition which was only beginning to develop in the early part of the twentieth century. Rose was one of three Americans who served on the League of Nations' health committee in 1935 to study the physiological bases of nutrition. She was a member of the American

Medical Association's Council on Foods. In 1937–1938, she was president of the American Institute of Nutrition (now ASN), which she helped found.

Awards established in her honor include a fellowship in her name awarded by the American Dietetic Association for graduate study in nutrition or allied fields. The ASN jointly with the Council for Responsible Nutrition has two Mary Swartz Rose Awards given to senior and junior investigators for outstanding preclinical and/or clinical research on the safety and efficacy of dietary supplements as well as essential nutrients and other biologically active food components that might be distributed as supplements or components in functional foods [3, 15, 21].

4.10   Ida Helen Ogilvie: Geologist (1874–1963)

Although Ida Helen Ogilvie (Fig. 4.8) came from a wealthy family, she was determined to be other than a debutante and wife. Ogilvie found her life's calling at Bryn Mawr where she studied geology under Florence Bascom, who had recently founded that institution's geology program. After graduating from Bryn Mawr, Ogilvie

Fig. 4.8 Ida Helen Ogilvie. (Courtesy of Wikipedia)

studied at the University of Chicago where she published her first paper in 1902. She completed her PhD at Columbia University. Her interests were glacial geography and petrology.

Ogilvie stayed at Columbia becoming the first lecturer in geology at Columbia University's Barnard College and teaching graduate geology courses at Columbia. From 1903 until her retirement in 1941, Ogilvie helped make geology careers accessible to women. She chaired Barnard's geology department, taught courses, and conducted research. Her research areas included glaciation in Canada and volcanic activities. She was the second woman admitted to the Geological Society of America.

Ogilvie also raised cattle, horses, dogs, and ponies at her farm in Bedford, New York. It became a model farm during World War I, "Bedford Camp," the first unit of the Women's Land Army in the U.S. She recruited young women into agriculture from around the U.S. After the war, she bought a larger farm and continued to breed her cattle. Ogilvie also helped endow scholarships for women in the sciences at Barnard, Bryn Mawr and Columbia [3, 15, 22].

4.11   Lise Meitner: Physicist (1878–1968)

An early pioneer in radioactivity (and called the German Marie Curie by Albert Einstein), Lise Meitner (Fig. 4.9) discovered protactinium with radiochemist Otto Hahn. She also played a major role in the discovery of atomic fission, which made atomic power as well as atomic weapons possible.

Raised in Vienna, Meitner developed an interest in physics at a young age but her father insisted that she study French so that she could support herself as a teacher if the need arose. Prevented from entering any Viennese high school because she was female, Meitner studied privately for the university examination. She completed eight years of schoolwork in two years and entered the University of Vienna in 1901. In 1906, she was the second woman to receive a doctorate in science from the university with a dissertation on heat conduction in non-homogeneous materials.

After her graduation, she was introduced to radioactivity and designed one of the first experiments demonstrating certain properties of alpha rays. She went to Berlin to study quantum physics with Max Planck and stayed for thirty-one years. She went to work with Otto Hahn, who was working on the chemistry of radioactivity at the Chemical Institute. Hahn helped Meitner set up a laboratory in the basement as Emil Fischer, who ran the laboratory that Hahn worked in, prohibited women from working in his laboratory. Meitner performed experiments in the cellar, being careful never to be seen upstairs. Prohibited from using the bathroom in the chemistry building, she used facilities at a nearby hotel. Sometimes she listened to lectures by hiding under a seat in the amphitheater.

By 1912, Meitner was receiving a small stipend and was provided with an assistantship at the Kaiser Wilhelm Institute for Chemistry. Her low income, however, meant she often was able to afford only black bread to eat and coffee to drink. In

Fig. 4.9 Lise Meitner.
(Courtesy of Smithsonian
Institution Archives)

1914, the University of Prague offered her a job, and the Kaiser Wilhelm Institute decided to pay her a salary. World War I interrupted Hahn's and Meitner's work together although Meitner continued much of it alone. By 1918, Meitner and Hahn had found the element protactinium.

In 1922, Meitner was allowed to lecture at the University of Berlin for the first time. In 1926, she was appointed "extraordinary professor," Germany's first woman professor of physics. Her laboratory work focused on clarifying the relationships between the beta spectra and gamma rays emitted by radioactive materials. In 1928, she received the Ellen Richards Prize (billed as the Nobel Prize for Women) from the Association to Aid Women in Science. In the 1930s, she began working with Hahn again to examine new heavy isotopes. He brought Fritz Strassmann to the team that included Dr. Clara Lieber, an American chemist, and a technician, Irmgard Bohne. However, in the 1930s, because of Nazi racial laws, Meitner was forced to flee Berlin leaving all of her personal belongings and scientific papers behind.

Meitner eventually went to work at the Nobel Institute in Stockholm, Sweden where she continued collaborating long distance with Hahn. Meitner's most important scientific contribution occurred here when she proposed that neutrons split uranium, causing decay products that were isotopes of barium. Meitner had explained nuclear fission and communicated this with Hahn. However, Hahn published his paper without acknowledging her contributions and received the Nobel Prize in

Chemistry in 1944 alone for the discovery of fission of heavy nuclei. The element, meitnerium, was named in her honor by German physicists in 1992 [15, 17, 23].

4.12   Ruth Florence Allen: Plant Pathologist (1879–1963)

Although Ruth Florence Allen was the "most cited woman in the past 30 years, mainly for her cytological studies of rust infections [in wheat]," that did not stop her from enduring a 50% pay cut during the Great Depression years of 1933–1936. The men that she worked with were not similarly impacted.[6]

Ruth Allen grew up in Wisconsin and her undergraduate, graduate and PhD degrees were earned at the University of Wisconsin. Her dissertation was on spermatogenesis and apogamy in ferns. She went to work for the U.S. Department of Agriculture combined with a position in the genetics department at the University of California, Berkeley. She retired in 1938 due to a variety of health issues.

But Ruth Allen made lemons out of lemonade. After her retirement, she played the stock market and built up a sizable fortune. After her death, her heirs donated a significant portion of those funds to the American Phytopathological Society for the Ruth Allen Award. Given every year since 1966, the award recognizes individuals who have made an outstanding, innovative research contribution that has changed, or has the potential to change, the direction of research in any field of plant pathology [1, 3, 24].

4.13   Amalie Emmy Noether: Mathematician (1882–1935)

A mathematician who was eulogized by Albert Einstein as "the most significant creative mathematical genius thus far produced since the higher education of women began," Emmy Noether (Fig. 4.10) was born and educated in Germany, receiving her PhD with highest honors at the University of Erlangen in 1907. Generally considered one of the greatest mathematicians of the twentieth century, her work in abstract algebra inspired so many mathematicians that a "Noether School" of mathematics is often referenced.

In 1918, she proved two theorems that formed a cornerstone for general relativity. One, now known as Noether's Theorem, established the equivalence between an invariance property and a conservation law. All of this while she was employed without pay at Göttingen University where women were not admitted to the faculty!

Noether emigrated to the U.S. in 1933 because being female and Jewish in Nazi Germany resulted in her being dismissed from her position as a faculty member at

---

[6] Her obituary read: "During the economy drives of the depression years between 1933–1936, her salary was reduced by half. Many considered this an unjustified discrimination against women in science, but Miss Allen accepted it as inevitable and without complaint."

Fig. 4.10 Emmy Noether.
(Courtesy Wikipedia)

the University of Göttingen. She accepted a position at Bryn Mawr College partly due to that institution's reputation for eminent women mathematicians. Unfortunately, she died of cancer within two years of her arrival in the U.S.

She was one of the earliest figures in twentieth-century theoretical physics and changed the appearance of algebra by her work on abstract algebra. Her name lives on through a mathematical structure called Noetherian rings, studied by many graduate students in mathematics. In 1982, the Emmy Noether Gymnasium, a coeducational school emphasizing mathematics, the natural sciences, and modern languages, opened in Erlangen [15, 16, 23, 25].

4.14   Ruth Benedict: Anthropologist (1887–1948)

Ruth Benedict (Fig. 4.11) was America's first woman anthropologist, the leading specialist in the field, and a mentor for well-known anthropologist, Margaret Mead. Benedict graduated Phi Beta Kappa from Vassar College with a degree in English literature and taught and did charity work before marrying Stanley Rossiter Benedict in 1914. Benedict enrolled at Columbia University's New School for Social

Research in 1919. There she discovered her interest in the field of anthropology and received her PhD in 1923.

Benedict's field work, very important to the education of an anthropologist, involved observing Native American tribes in the summer. She observed the Serrano, Zuni, Pirma, Mescalaro Apache, and Blackfoot. She served as a lecturer at Columbia from 1923 until 1930 with almost no pay because she was a married woman. In 1931, she was appointed assistant professor. By this time, she and her husband were separated, and her pay was increased. *Patterns of Culture*, her groundbreaking book, was published in 1934, and was used as an introduction to anthropology textbook for the next 25 years. Also in 1934, she was named a fellow at the New York Academy of Sciences. In 1936, she was promoted to associate professor and in 1937 she became the executive officer of the anthropology department.

Benedict worked for the Office of War Information from 1943 to 1945. She took a leave of absence from Columbia to write an analysis of Japanese culture, *The Chrysanthemum and the Sword*, which became another major landmark of American anthropology. In 1947, the U.S. government awarded her a large grant to establish and direct a program of research into contemporary cultures. And, in 1948, just a few months before her death, Columbia made her a full professor. Benedict was inducted posthumously into the National Women's Hall of Fame [10, 15].

4.15   Gerty Cori: Biochemist (1896–1957)

Nobel Laureate Gerty Cori (Fig. 4.12) was the first American woman to win a Nobel Prize in science. She and her husband Dr. Carl Ferdinand Cori shared the 1947 Nobel Prize in Physiology or Medicine with Bernardo Alberto Houssay "for their discovery of the course of the catalytic conversion of glycogen." They explained the physiological process by which the body metabolizes sugar.

Cori was born in Prague where her uncle, a professor in pediatrics, nurtured her interest in mathematics and science and encouraged her to undertake the study necessary to enter a university and study medicine. By age 18, she had passed a very difficult examination and entered the German branch of the medical school at Prague's Carl Ferdinand University. During her first semester anatomy class, she met her husband-to-be. They jointly agreed to pursue medical research, not medical practice, and to jointly attain medical certification (a six-year process) before marrying.

Fig. 4.12 Gerty and Carl Cori. (Courtesy Smithsonian Institution Archives)

In 1922, Carl received an offer to work in the U.S. and Gerty, demonstrating significant independence, stayed behind until she too had an offer to work in the U.S. They both worked at the New York State Institute for the Study of Malignant Diseases in Buffalo (later the Roswell Park Memorial Institute), New York; he as a biochemist, she as an assistant pathologist. Here, Gerty experienced resistance to her presence as a woman in science. The director of the Institute threatened to fire Gerty if she did not end collaborative work with her husband. Later, a university offered Carl a job – only if he ended working collaboratively with his wife. The rationale for these requests was that not only was it un-American for a man to work with his wife – his wife was standing in the way of his career advancement!

Not everyone believed this, however. After becoming naturalized American citizens in 1928, Gerty and Carl received offers to work at Washington University in St. Louis. Carl would become a professor of pharmacology and Gerty was offered the position of research associate in pharmacology. Here, Gerty gave birth to their son, Thomas who eventually became a research chemist himself, following in his parents' footsteps.

Although denied positions and titles that she would have received as a man, Gerty was promoted to associate professor in biochemistry in 1943, the year she and Carl achieved the synthesis of glycogen in a test tube. In 1947, shortly before she was awarded the Nobel Prize, Gerty was promoted to full professor of biochemistry. The Cori's discovery of glycogen led to more effective treatments for diabetes. The relationships between the liver and muscle glycogen, and blood glucose and lactic acid is now known as the Cori cycle. Gerty's other areas of research included hereditary glycogen storage diseases in children and the identification of a new enzyme, amylo-1, 6-glucopsidase which helped her identify the structure of glycogen. She became a member of the National Academy of Sciences in 1948. Posthumously, she was inducted into the National Women's Hall of Fame [15–17, 23, 26].

4.16   Irene Joliot-Curie: Chemist and Physicist (1897–1956)

The eldest daughter of famed scientists, Marie and Pierre Curie, Irene Joliot-Curie was in her own right a significant scientist. And, like her mother, she was the recipient of the Nobel Prize in Chemistry. She shared that 1935 Nobel Prize with her husband Frédéric Joliot-Curie (Fig. 4.13), for the discovery of artificial radioactivity.

Irene's upbringing was influenced by the untimely death of her father and her close association with her paternal grandfather, Eugene Curie, who taught her botany and natural history, but also shaped Irene's left-leaning political sentiments and aversion to religion. Irene studied at the Radium Institute in Paris, founded by her parents, and eventually succeeded Marie Curie as the research director. She met her husband, Frédéric Joliot, there, who insisted that they publish and be known by the combined name Joliot-Curie so as not to lose the Curie name – Marie and Pierre had two daughters – Irene and Eve.

Fig. 4.13 Irene and Frédéric Joliot-Curie. (Courtesy of Wikipedia)

Irene accompanied Marie to the front lines in World War I where Marie used the new x-ray equipment to treat soldiers. By age 21, she had become her mother's assistant at the Radium Institute. In 1925, Irene completed her doctoral thesis on the emission of alpha rays from polonium – an element that her parents had discovered. She and Frédéric used alpha particles to bombard aluminum nuclei and other elements and discovered artificial radioactivity. Their results were presented to the Academy of Sciences in 1934.

Artificial radioactivity has many applications including a wide range of medical diagnoses. The award of the Nobel Prize was bittersweet for Irene, as the announcement came several months after Marie Curie had died [15, 23].

4.17 Katharine Burr Blodgett: Physicist (1898–1979)

The first woman scientist at General Electric (GE), Katharine Blodgett (Fig. 4.14) was the inventor of invisible, or non-reflecting, glass. This glass is used extensively in camera and optical equipment. In fact, one of its first applications was in a projection lens for the 1939 movie *Gone With the Wind*. The surface chemistry techniques that she developed with her mentor, Irving Langmuir, are called Langmuir-Blodgett films.

Blodgett won a scholarship to attend Bryn Mawr, where she was became intrigued with math and physics, and graduated second in her class. She wanted to work at GE, as had her father, but Irving Langmuir, 1932 Nobel Laureate in Chemistry, advised her first to broaden her scientific education. She received her master's degree from the University of Chicago where her thesis topic had been

Fig. 4.14   Katharine Blodgett. (Courtesy of Smithsonian Institution)

influenced by World War I and the German's use of poisonous gas. She studied the adsorption of gases by coconut charcoal and thus improved the chemical structure of gas masks.

Blodgett then went to work for GE at a time when it was almost impossible for women to get professional level jobs in corporations. With Langmuir's support, she studied at Cambridge University under Nobel Laureate Sir Ernest Rutherford and became the first woman to earn a doctorate at Cambridge University in 1926.

In 1938, Blodgett announced her invention of non-reflecting glass. This glass is now used in automobile windows, showcases, eyeglasses, picture frames, and submarine periscopes, in addition to cameras and optical equipment. During World War II, she worked on plane wing deicing and invented a smoke screen that saved many lives in campaigns in North Africa and Italy. In 1947, she invented an instrument to measure humidity in the upper atmosphere using weather balloons.

Blodgett received awards and honorary degrees for her inventions. She was the first industrial scientist to win the Francis P. Garvan Medal (1951) given by the American Chemical Society to honor American women for distinguished service in chemistry. She was posthumously inducted into the National Inventors Hall of Fame [15, 17].

References

1. M.W. Rossiter, *Women Scientists in America: Struggles and Strategies to 1940*, vol 1992 (The Johns Hopkins University Press, Baltimore)
2. E. Flexner, E. Fitzpatrick, *Century of Struggle: The Women's Rights Movement in the United States*, Enlarged edn. (The Belknap Press of Harvard University, Cambridge, 1996)
3. M. Ogilvie, J. Harvey (eds.), *The Biographical Dictionary of Women in Science: Pioneering Lives from Ancient Times to the Mid-twentieth Century* (Routledge, New York, 2000)
4. H.C. Hoerle, F.B. Saltzberg, *The Girl and The Job* (Henry Holt and Company, New York, 1919)
5. O. L. Hatcher (ed.), *Occupations for Women: A Study Made for the Southern Woman's Educational Alliance* (L.H. Jenkins, Inc, Richmond, 1927)
6. W.K. LeBold, D.J. LeBold, Women engineers: a historical perspective. ASEE Prism **7**, 30–32 (1998)
7. D.E. Fleischman, *An Outline of Careers for Women: A Practical Guide to Achievement* (Doubleday, Doran and Company, Inc., New York, 1929)
8. M.S. Leuck, *Fields of Work for Women* (D. Appleton-Century, Company, New York, 1938)
9. M.W. Rossiter, *Women Scientists in America: Before Affirmative Action 1940–1972*, vol 1995 (The Johns Hopkins University Press, Baltimore)
10. B. Harris, *Beyond her Sphere: Women in the Professions in American History* (Greenwood Press, Westport, 1978)
11. A.M. Barker, Women in Engineering During World War II: A Taste of Victory, 21 Nov 1994, Rochester Institute of Technology (unpublished)
12. A.S. Bix, 'Engineeresses' 'Invade' campus: four decades of debate over technical coeducation. In: Proceedings of the 1999 International symposium on technology and society – women and technology: historical, societal, and professional perspectives, pp. 195–201, New Brunswick, 29–31 July, 1999
13. U.S. Department of Labor, Women's Bureau, *The Outlook for Women in Science*, Bulletin of the Women's Bureau No. 223-1, 1949
14. B. F. Shearer, B. S. Shearer (eds.), *Notable Women in the Life Sciences* (Greenwood Press, Westport, 1996)
15. P. Proffitt (ed.), *Notable Women Scientists* (The Gale Group, Detroit, 1999)
16. M.J. Bailey, *American Women in Science: A Biographical Dictionary* (ABC-CLIO, Denver, 1994)
17. B. F. Shearer, B. S. Shearer (eds.), *Notable Women in the Physical Sciences* (Greenwood Press, Westport, 1997)
18. M.B. Ogilvie, *Women in Science: Antiquity through the Nineteenth Century, a Biographical Dictionary with Annotated Bibliography* (MIT Press, Cambridge, 1993)
19. P.J. Read, B.L. Witlieb, *The Book of Women's Firsts* (Random House, New York, 1992)
20. A.C. Goff, *Women Can Be Engineers* (Edwards Brothers, Inc., Ann Arbor, 1946)
21. Mary Swartz Rose Awards Honor Leaders in Nutrition Research. https://www.nutraceuticals-world.com/issues/2013-06/view_people-in-the-news/mary-swartz-rose-awards-honor-leaders-in-nutrition-research, 24 Apr 2013. Accessed 6 Apr 2020
22. Women's Land Army of World War I. https://www.womenshistory.org/resources/general/womens-land-army-world-war-i. Accessed 6 Apr 2020
23. S.B. McGrayne, *Nobel Prize Women in Science: Their Lives, Struggles, and Momentous Discoveries* (Carol Publishing Group, New York, 1993)
24. Ruth Allen Award. https://www.apsnet.org/members/give-awards/awards/RuthAllen/Pages/default.aspx. Accessed 6 Apr 2020
25. C. Morrow, T. Perl (eds.), *Notable Women in Mathematics: A Biographical Dictionary* (Greenwood Press, Westport, 1998)
26. G. Kass-Simon, in *Women of Science: Righting the Record*, ed. by P. Farnes, (Indian University Press, Bloomington, 1990)

Chapter 5
Sputnik Signals a New Kind of War

5.1 The Korean War

With the start of the Cold War after World War II, and then the Korean War in 1950, American women were once again asked to contribute to the nation's defense. Young women were even encouraged to study science and engineering. In 1951, President Harry Truman was seeking a standing army of 3.5 million men and highly trained human resources at home – scientists and engineers – not only for Korea but anywhere necessary for the foreseeable future. The shortage of these resources was especially acute because of low birth rates during the 1930s and the drop off in enrollments and employment after World War II [1, 2].

The Office of Defense Mobilization (ODM), in its Defense Manpower Policy No. 8 (September 1952), published a significant policy statement advocated by Arthur Flemming, assistant to the director for manpower of ODM and a strong supporter of women [1]:

Throughout this document all references to scientists and engineers make no distinction between the sexes or between racial groups; it being understood that equality of opportunity to make maximum effective use of intellect and ability is a basic concept of democracy.

In addition, the policy's eleventh of twelve recommendations was for employers of scientists and engineers "to reexamine their personnel policies and effect any changes necessary to assure full utilization of women and members of minority groups having scientific and engineering training." Flemming was expressing what was increasingly becoming the official governmental view: that women were needed as scientists and engineers. However, full and equal opportunity for women in the scientific fields had yet to be realized as the Committee of Specialized Personnel from ODM reported on December 9, 1953 [1, 3]:

For the most part, the female graduate [i.e., in engineering and the sciences] has been held down as far as advance in classification and remuneration is concerned.

© The Editor(s) (if applicable) and The Author(s), under exclusive license to
Springer Nature Switzerland AG 2020
J. S. Tietjen, *Scientific Women*, Women in Engineering and Science,
https://doi.org/10.1007/978-3-030-51445-7_5

Such action on the part of management is totally unrealistic, and in order to pro-mote the development of our high potential of female scientists and engineers, this unrealistic sex barrier must be broken.

The federal government's official policy throughout the 1950s was to encourage women to enter scientific and technical fields and to urge employers to hire them and utilize them fully (including the federal government itself). However, no federal incentives, such as tax credits, for fuller utilization of womanpower or enforcement mechanisms were put in place [1].

The pendulum toward encouragement of women to be engineers and scientists had swung again. The country needs women to be in the workforce and supporting the war effort when the country is at war, but then willing and compliant about being discarded and being replaced at the end of periods of national crisis. After the mid-dle of the twentieth century, with higher levels of education and training among women and the general population, such treatment wasn't going to be acceptable for too much longer. But it would still be several decades before significant progress was made toward anything that looked like more equal opportunity for scientific and engineering women.

5.2   Off to Suburbia

Women in the U.S. in the 1950s were being pulled in two directions at the same time. The average age of marriage for American women dropped significantly dur-ing the period 1945–1950. The birth rate soared, especially among the college edu-cated (the children born in this period were called the baby boomers). Marriage took precedence over careers. In addition, a mass white exodus to suburbia began and for the first time, college-educated, middle class women had as many children as poor women did [4]. Television and advertising glorified domesticity and the housewife, especially her role as a consumer. Yet, in the increasingly consumer-based economy, more workers were required to design and produce all of the new products. Thus, there were many more economic opportunities for women in the workforce [5].

In the face of declining female enrollments in science and engineering and with the projected shortages of technical manpower, in April 1956, President Eisenhower, with the urging of the Office of Defense Mobilization Director Arthur Flemming established a National (later called President's) Committee on the Development of Scientists and Engineers to serve as a clearinghouse for the many nongovernmental efforts being undertaken around the country to train more scientists and engineers. Interestingly, nineteen men and no women were appointed to this committee and its vice chairman seemed particularly uninterested in recruiting women [1]. The Committee's Second Interim Report to the President on October 4, 1957, included the following [6]:

...Obviously, steps must be taken to break down employment barriers to women in science, engineering, and the technician fields. Public education programs of many varieties are needed to encourage young women to undertake science and engineering studies and to ensure that they receive satisfactory employment after training. Employment requirements

and specifications, job content, employment conditions, and environment need to be recon-
sidered. Long established prejudices against women in engineering and science need to be
broken down not only among employers, supervisors, and co-workers but among the women
themselves.

In its publications and public statements, the Committee has pointed out the advantage,
and indeed, the necessity of developing the full potential of the Nation's womanpower qual-
ified for scientific and technological pursuits.

The committee was disbanded in December 1958. However, the vice chairman's request for a breakdown of data into gender would prove beneficial later for female members of the National Science Foundation's (NSF) Divisional Committee for Scientific Personnel and Education [1].

5.3 Sputnik Is Launched

With the launching of Sputnik in October 1957, Americans began to focus their anti-Communist sentiment on the scientific and education arena. Scientists had begun trying to increase funding and emphasize scientific education earlier in the 1950s as the Cold War intensified. When Sputnik went up, Soviet superiority in science was made quite visible to the American public. And there was a new tone of urgency and stridency associated with the talk about the need to recruit women scientists and engineers [1, 7].

In response, President Eisenhower exhorted the American people to meet the need for thousands of new scientists, saying "this [national security] is for the American people the most critical problem of all ... we need scientists by the thousands more than we are now presently planning to have." Further, the President requested that the NSF "develop a program for collection of needed supply, demand, employment and compensation data with respect to scientists and engineers" [7].

The NSF accomplished this request through its Scientific Manpower Program and that program's two elements, Manpower Studies and the National Register. This National Register of Scientific and Technical Personnel grew out of the National Roster of World War II and subsequent efforts aimed at Cold War preparedness. Data for the National Register had been collected as early as 1954 but little was published prior to 1959 when in response to the Sputnik launch, Congress increased the NSF's budget. Consequently much of the data available have significant gaps and data on women scientists and engineers is particularly incomplete [1, 7].

The NSF designed programs to provide federal assistance to the "best and brightest" in order to produce the scientists needed for the future and to gather the necessary data. While Congress discussed science budgets and fellowship programs as part of the U.S. response to the Sputnik launch (with training scientists and engineers now a matter of national survival), articles in such a prestigious newspaper as *The Wall Street Journal* titled "Science Talent Hunt Faces Stiff Obstacle: 'Feminine Fallout'; Officials Fear Many Federal Scholarships Will Go to Girls – Who'll Shun Careers" appeared. Because up to one-third of these fellowships were expected to go to women who would marry, have children, and interrupt their careers, the author commented:

Hence it's inevitable that some Government money will go to train scientists who experiment only with different household detergents and mathematicians who confine their work to adding up grocery bills.

But, the author further lamented, it would not be feasible to place quotas on the number of fellowships given to women as this "probably would embroil the Government in a great controversy with the many 'equal rights' advocates among the ladies" [1, 7].

The National Defense Education Act (NDEA) was finally passed by Congress in 1958. The act clearly linked higher education to national defense by declaring [1]:

The Congress hereby finds and declares that the security of the Nation requires the fullest development of the mental resources and technical skills of its young men and women ...

We must increase our efforts to identify and educate more of the talent of our Nation. This requires programs that will give assurance that no student of ability will be denied an opportunity for higher education because of financial need; will correct as rapidly as possible the existing imbalances in our educational programs which have led to an insufficient proportion of our population educated in science, mathematics, and modern foreign languages and trained in technology.

Ten new programs were established upon enactment of the NDEA including a federal student loan program and a new graduate fellowship program larger and broader than the one at the NSF. These fellowships would continue until 1973 [1].

However, women were still feeling a conflict between their domestic obligations and pursuing a scientific or engineering career. And now with a perceived patriotic duty, especially at a time when recruitment literature stressed that Russian women constituted about half of the combined scientific and engineering workforce in that country, articles on both sides of the issue appeared in popular magazines with such titles as "Woman's Place Is in the Lab, too," "Science for the Masses," "Bright Girls: What Place in Society?" "Plight of the Intellectual Girl," and "Female-Ism: New and Insidious" [1, 7]. The Women's Bureau of the Department of Labor stated "Clearly, the Nation needs qualified women scientists, and those women who are interested and have the capacity should be encouraged to consider careers in scientific fields" [6]. The Women's Bureau further observed:

Many women have personal qualities of special value for scientific work. Among these are an inquiring mind; imagination; and a penchant for detail, for orderly, logical thinking, for precise description and measurement, and for critical analysis of facts and theories. The basic requirements common to all scientific fields are interest and a mental capacity that can be developed through specialized training to solve problems and search for a deeper understanding of the nature of things. [6]

Its tone for the future of women was completely optimistic. Under the heading of "Women Have the Abilities," the Women's Bureau opined [6]:

The achievements of women in every field of science amply demonstrate women's abilities for scientific endeavors. The psychologists tell us that little, if any, difference exists in general intelligence between men and women. Test results of the General Aptitude Test Battery, developed by the U.S. Employment Service and used widely with high school seniors, also bear out the fact that there are no significant differences in aptitudes between boys and girls, and also that there are greater differences among individual girls and among individual boys than between girls as a group and boys as a group.

> *Moreover, what differences exist may be attributed largely to different experiences and backgrounds. The National Manpower Council's report on "Womanpower" goes beyond this and points out that there is some evidence "that women studying mechanical and technical subjects receive better grades than men who score equally well on aptitude tests." An observation of a college professor of physics provides still further evidence of women's basic aptitudes for scientific work. He states that differences in mechanical and electrical background are largely overcome by women through additional laboratory work in their freshman year; and that by the sophomore year they are at least as competent as men in the theoretical aspects of science.*

Yet the number of women receiving degrees in the physical sciences was still relatively small and by 1957–1958 had not yet risen back to the level observed in 1947–1948 (Table 5.1) [6].

The first comprehensive study describing U.S. scientific and technical manpower was published in 1964 – it did not examine traits such as sex and ethnicity [7]. The NSF also funded a number of studies to identify the factors associated with the low numbers of women in science and engineering. These studies showed that myths about women not being suited for the sciences due to ability, emotion, or motivation were just that – myths – and the studies recommended actions to encourage women to pursue scientific careers [5].

Post-Sputnik, the major cultural and legislative changes of the 1960s would set the stage for greater numbers of women scientists by the turn of the century.

5.4 1960s Activism

The women's movement experienced a dramatic rebirth in the 1960s that later translated into significantly increased professional opportunities for women. That it occurred at the same time as the civil rights movement is probably much less of a coincidence than it appears. The birth of feminism and the demand for women's

Table 5.1 Degrees granted to women in the physical sciences 1947–1948 through 1957–1958 (all levels)

Academic year	Total degrees	Percent change from previous year
1957–1958	2006	+5%
1956–1957	1905	+6%
1955–1956	1789	+9%
1954–1955	1641	+9%
1953–1954	1511	+1%
1952–1953	1503	−7%
1951–1952	1613	−6%
1950–1951	1710	−17%
1949–1950	2051	−13%
1948–1949	2344	−13%
1947–1948	2696	–

suffrage in the 1800s had been closely aligned with the abolitionist cause. Now, the rebirth of the women's movement was closely related to the struggle for racial equality. Indeed, the militancy of college students during the 1960s mirrored some of the rebellious activism that had been effective and prominent during the suffrage movement. In the 1960s, protests were held on campuses and in the streets, and students traveled to the South on behalf of civil rights [4].

Betty Friedan's *Feminine Mystique*, published in 1963, launched an attack on suburban America and the status and roles assigned to women. Friedan meant her book as a call to action, and indeed, many women strengthened their resolve to take charge of their own lives as a result of its publication. The percentage of college-educated females who worked outside the home increased from 7% in 1950 to 25% in 1960 as it became apparent to many such women that housework was not their calling [4].

In the 1960s, corrective legislation that addressed women's historically lower status in society relative to men began to roll out, one after the other. And by 1962, 53% of all female college graduates were employed, while 36% of those with high school diplomas held jobs. Seventy percent of all females who had five or more years of higher education worked [4].

5.4.1 Presidential Commission on the Status of Women

The Presidential Commission on the Status of Women was convened in 1961 to investigate and suggest remedies for "prejudices and outmoded customs [that] act as barriers to the full realization of women's basic rights." Seven committees representing various facets of American life – civil and political rights, education, federal employment, private employment, home and community, social security and taxes, and protective labor legislation – were involved in the commission's work. Their final report, issued in 1963, proved that in almost every area, women were second-class citizens. President Kennedy took two actions as a result of the work that went into the commission's report:

1. Women were to be on an equal basis with men for Civil Service promotion
2. All executive department promotions were to be based on merit [1, 4, 8]

5.4.2 Equal Pay Act

After the publication of the report from this Commission and in large part because of its findings, President Kennedy signed the Equal Pay Act (Fig. 5.1), which states that "… no employer shall discriminate between employees on the basis of sex by paying wages for equal work, the performance of which requires equal skill, effort and responsibility, and which are performed under similar working conditions." The act was sponsored by Edith Green, of Oregon, one of the most influential members

Fig. 5.1 President Kennedy signing the Equal Pay Act on June 10, 1963. (Courtesy of John F. Kennedy Presidential Library and Museum)

of Congress.[1] It was the first major piece of legislation addressing sexual inequality since the Nineteenth Amendment. Although there were significant exemptions included as part of the Act, the legislation was an important step forward [4, 8, 9].

5.4.3 Civil Rights Act: Title VII

In 1964, a second major piece of legislation – Title VII of the Civil Rights Act – was passed to prohibit discrimination in employment on the basis of race, religion, color, national origin, and sex. The original intent of the bill was to deal with racial

---

[1] Edith Green (1910–1987) served as a U.S. Representative from Oregon from 1955–1974. She left a substantial legacy in the U.S. Congress. She impacted almost every education bill enacted during her tenure. Green supported federal aid to education and the anti-poverty programs of the Great Society while resisting expansion of the federal bureaucracy. She was appointed to the Committee on Education and Labor in her freshman term in the House of Representatives where she served until her final term in the House when she took a seat on the Committee on Appropriations. As chair of the Education and Labor subcommittee on higher education, she was responsible for establishing the first federal program for undergraduate scholarships.

inequality. The amendment adding the word "sex" was proposed by the powerful chair of the House Rules Committee, Howard Smith of Virginia, in an effort to retard its passage. Smith urged Congress "to protect our spinster friends in their 'right' to a husband and family," a conniving plea that was met with roars of laughter. His apparent intent was to burden the entire law with the addition of gender and cause its defeat due to the expected ensuing controversy and ridicule. Thereafter, his strategy of adding "sex" was referred to as a "joke." Nonetheless, the amendment to the language was retained, and the law passed. The Equal Employment Opportunity Commission was formed to enforce Title VII and found that most of its complaints were from women, not, as had been expected, from minorities [4, 10].

5.4.4   The Dawn of Affirmative Action

In September 1965, President Johnson essentially began affirmative action by signing Executive Order 11246. This order required all companies wishing to do business with the federal government to not only provide equal opportunity for all but also to take affirmative action – defined as extra steps – to bring their hiring in line with available labor pools by race [11].

5.4.5   Recognition of Sex Discrimination

Two years later, in 1967, President Johnson signed Executive Order 11375 extending Executive Order 11246 to include "sex" as a protected category. This executive order now required that affirmative action be taken on behalf of women in addition to minorities, as required by Executive Order 11246, so that hiring was in line with gender proportions as well as racial proportions in the relevant labor pools [11].

5.4.6   National Organization for Women

In 1966, the National Organization for Women was founded. NOW describes itself as a civil rights organization to bring women into "truly equal partnership with men in all areas of American society." The NOW 1968 Bill of Rights called for support of a wide array of issues including equal and unsegregated education and equal job-training opportunities [4].

By the end of the 1960s, the U.S. had successfully landed men on the moon, symbolizing American technical and scientific superiority over the Soviet Union. The women's rights and civil rights movements encouraged women and minorities to pursue all career fields – including nontraditional ones – although the number of women earning PhDs in 1969 in the physical sciences totaled 186, just 5.4% of the total. However, by the early to mid-1970s, the Vietnam War, the energy crisis, and a widening awareness of environmental issues somewhat soured Americans on science and technology. Now, scientists were needed to help save America from themselves – so maybe women and minorities, with a new and different way of solving problems – could help [7, 12].

***

KEY WOMEN OF THIS PERIOD

***

5.5   Elizabeth Lee Hazen: Microbiologist and Mycologist[2] (1885–1975)

One of the first three women to be inducted into the National Inventors Hall of Fame (with Rachel Fuller Brown), Elizabeth Lee Hazen (Fig. 5.2) co-discovered nystatin, "the first broadly effective antifungal antibiotic available to the medical profession". Hazen was born in Mississippi to cotton farmers and was orphaned at age three. Raised by an aunt and uncle, Hazen graduated from what is today Mississippi University for Women with a BS and a certificate in dressmaking. She taught high school physics and biology and persevered with her studies, earning an MS from Columbia University in 1917. At age 42, she completed her PhD at Columbia in microbiology.

Fungus (or mycotic) infections became the source of her investigation as she was determined to find a naturally occurring antifungal antibiotic that would be safe for human use. In 1948, she collected a soil sample near Warrenton, Virginia from which she isolated a culture that yielded two antifungal antibiotics. One of these was too toxic to test on animals, but the second became "nystatin" named by E.R. Squibb and Sons and commercialized by Squibb after its 1957 patent. Awards and honors showered upon Hazen and Brown as nystatin (named for New York State, where they both worked) is highly effective for oral, vaginal, and skin infections caused by *Candida* species.

Brown and Hazen used the royalties they received over the 17-year life of the nystatin patent to establish a research fund. The Brown-Hazen Fund provided grants to support basic research in biochemistry, microbiology and immunology. The Fund allocated scholarships in the inventors' names to the women's colleges they had each attended and supported programs advancing women's participation in the sciences. Its monies lasted until 1976. Hazen's estate provided funds for scholarships at the Mississippi University for Women, matched by the Brown-Hazen Committee, that are awarded to this day. The University renamed one of its laboratories the Elizabeth Lee Hazen Microbiology Laboratories [13, 14].

---

[2] Mycology is the study of fungi.

Fig. 5.2 Elizabeth Lee Hazen (left) and Rachel Fuller Brown (right). (Courtesy of Smithsonian Institution)

5.6   Edith H. Quimby: Biophysicist (1891–1982)

Edith Quimby pioneered diagnostic and therapeutic applications for x-rays, radium, and radioactive isotopes in the early days of radiology. Her research allowed physicians to calculate the dosage needed with the fewest side effects. She also worked to protect workers who handled the materials from harm.

Quimby's curiosity in science was encouraged by her father and challenged by her high school science teacher. She became the first woman to take the mathematics/physics major at Whitman College and graduated with a BS in 1912. A teaching fellowship in physics at the University of California enabled her to work toward a masters degree in physics. She also met her husband, physicist Shirley Quimby, at the University of California whom she married in 1915. In 1917, she received her MA.

The Quimbys moved to New York for Shirley Quimby to work on his PhD. However, because his part-time instructorship was not enough to support them, Edith needed to find a job. In 1919, Edith Quimby went to work at the New York City Memorial Hospital for Cancer and Allied Diseases as an assistant physicist. She was one of the few women in America engaged in medical research and fortunate to have an employer, physicist Dr. Giocchino Failla, who did not object to hiring and working with women.

Radiation and radiological physics was still in its infancy and thus Edith Quimby became a pioneer in radiation physics. From 1920 to 1940, Quimby wrote 50 technical articles on radium and x-ray treatments and the correct dosages for cancer patients. Recognizing that this was the first practical guidance on the topic for physicians, she received the highest honor of the American Radium Society, the Janeway Medal, in 1940. She was the first woman to receive the medal and only the second person without an MD. In 1941, she received the Gold Medal of the Radiological Society of North America for her "continuous service to radiology." Marie Curie was the only other woman to have received this Gold Medal. Later, Quimby worked on the Manhattan Project and became an expert on procedures for cleaning up radioactive spills.

Quimby was made a full professor at Columbia University in 1954. There she established the Radiological Research Laboratory to study the medical uses of radioactive isotopes, including their application to cancer diagnosis and treatments. She also examined applications to thyroid disease and diagnosis of brain tumors, making her a pioneer in the field of nuclear medicine [13, 15].

5.7   Ida Tacke Noddack: Chemist (1896–1978)

Nominated unsuccessfully three times for the Nobel Prize in Chemistry, German chemist Ida Tacke Noddack co-discovered element 75, rhenium, named for the Rhine River. Noddack collaborated throughout her career with her husband, German chemist Walter Noddack. In 1934, Noddack disagreed with physicist Enrico Fermi[3] about the transuranic elements that he believed resulted from neutron bombardment of uranium; believing instead that he had discovered nuclear fission. This conclusion of hers was met with skepticism and she was not credited when nuclear fission was, in fact, announced in 1938.

After receiving her doctorate in engineering in 1921, Noddack was determined to fill in the holes in the periodic table, developed by Russian chemist Dmitri Ivanovich Mendeleev, for elements 43 and 75. These elements are located in Group VII under manganese. In 1925, she, Walter Noddack, and x-ray specialist Otto Berg were successful with element 75. They thought they had also discovered element 43, but that element is found to be produced only artificially.

She and Walter worked together until his death in 1960 and published more than one hundred scientific papers together. They received the Leibig Medal from the German Chemical Society for the discovery of rhenium. Ida Noddack received the High Service Cross of the German Federal Republic in 1966. She was an honorary member of the Spanish Society of Physics and Chemistry and received an honorary doctorate of science from the University of Hamburg [15, 16].

---

[3] Physicist Enrico Fermi created the world's first nuclear reactor. He received the 1938 Nobel Prize in Physics.

5.8   Florence Seibert: Biochemist (1897–1991)

Florence Seibert (Fig. 5.3) had contracted polio when she was three years old which caused her to limp for the rest of her life. As she read biographies of scientists as a teenager, her interest was sparked in pursuing a scientific career. She did pursue a scientific education and received her undergraduate degree in 1918 from Goucher College with emphases in chemistry and biology. Heeding the call for women scientists during World War I, she filled a position vacated by a man and coauthored papers on the chemistry of cellulose and wood pulps while working in the Chemistry Laboratory at Hammersley Paper Mill. In 1923, Seibert received her Ph.D. in biochemistry from Yale University.

Seibert identified the impurities introduced into intravenous injections and developed a new spray-catching trap to prevent contamination during the distillation process. This process was subsequently adopted by the Food and Drug Administration, the National Institutes of Health and pharmaceutical firms and was important in facilitating intravenous blood transfusions.

Seibert is best known for her work in identifying a pure form of tuberculin with her first paper published on the topic in 1934. Her work enabled the development and use of a reliable tuberculosis skin test that became the national and international standard for tuberculin tests in the 1940s. The recipient of many awards for her groundbreaking work, Seibert has been inducted into the National Women's Hall of Fame [15].

Fig. 5.3   Florence Seibert.
(Courtesy Smithsonian
Institution Archives)

5.9   Rachel Fuller Brown: Biochemist (1898–1980)

Along with Elizabeth Lee Hazen, Rachel Fuller Brown (Fig. 5.2) was among the first three women inducted into the National Inventors Hall of Fame in recognition of the development of nystatin, the most important biomedical breakthrough since the discovery of penicillin. Nystatin has cured sufferers of life-threatening fungal infections, vaginal yeast infections, and athlete's foot. The royalties earned over the 17-year life of the patent for nystatin were dedicated to scientific research through the Brown-Hazen Fund. The Fund is credited with helping unknown young scientists and significantly contributing to human health in medical mycology – the study of fungi. The Brown-Hazen Fund awarded grants over its lifetime and as the patent was ready to expire, the Fund Committee gave monies to Mount Holyoke which established a fund to sponsor Brown Fellows.

Brown attended Mount Holyoke, with her education financed by a wealthy friend of the family. She was going to be a history major, but discovered chemistry when she was fulfilling a science requirement. She earned her AB degree in history and chemistry in 1920 and completed her MA from the University of Chicago in organic chemistry. After teaching chemistry and physics in the Chicago area, she returned to the University of Chicago to complete her PhD in organic chemistry with a minor in bacteriology. She was finally awarded her doctorate in 1933, seven years after she submitted her thesis when she was able to return to Chicago from her job in New York to take her oral examinations.

Her early work at the Division of Laboratories and Research at the New York State Department of Health in Albany led to the development of a pneumonia vaccine that is still in use today. In 1948, she embarked on the project with Hazen that led to the discovery of an antifungal antibiotic to fight fungal infections. Their discovery, nystatin, was patented through the nonprofit Research Corporation. E.R. Squibb and Sons was able to develop a safe and effective method of mass production of the drug. Nystatin has been used in agricultural and livestock applications and has even been used to restore art works. In 1951, Brown was promoted to associate biochemist. By the time of her death in 1980, she had paid back the wealthy woman who had financed her education [14, 15].

5.10   Margaret Mead: Anthropologist (1901–1978)

The foremost anthropologist of the twentieth century, Margaret Mead (Fig. 5.4) brought an awareness of anthropology to the general public. Several of her books, including *Coming of Age in Samoa* (1928), *Sex and Temperament in Three Primitive Societies* (1935), and *Male and Female* (1949), became best sellers. Her expeditions to Samoa, New Guinea, Bali, and to Native American tribes provided material for more than 1500 books, articles, films, and occasional pieces. She popularized the idea that human differences arose not just from biological determinants but also from the mores and traits of individual cultures.

Mead was the first anthropologist to study how children grow up and to compare child-rearing practices among women from various cultures. She was the first to study women's roles in various cultures, as male anthropologists had largely ignored women and children in their studies. She founded the culture and personality school of anthropology that examines the way a culture shapes any individual's personality. She and her third husband, Gregory Bateson, used photography and then movies and video to record vanishing cultures. Her main areas of research included personality and culture; child development; and the application of psychoanalytic theory, learning theory, ethology, and cybernetics in studies of seven oceanic cultures.

Mead studied at Barnard College in New York City where she met Ruth Benedict, an instructor in the Anthropology Department under Franz Boas. Benedict was Boas' protégé and Mead would become Benedict's. Mead received her PhD from Columbia in anthropology in 1929. She would have the position of adjunct professor at Columbia for most of her career.

Mead was president of the American Anthropological Association in 1960 and the American Association for the Advancement of Science in 1975. She was elected a member of the National Academy of Sciences in 1975 and the American Philosophical Society in 1977. Mead has been inducted into the National Women's Hall of Fame and featured on a U.S. postage stamp. Posthumously, she received the Presidential Medal of Freedom [15, 16].

Fig. 5.4 Margaret Mead. (Courtesy Library of Congress)

5.11   Barbara McClintock: Geneticist (1902–1992)

Barbara McClintock (Fig. 5.5) received the Nobel Prize in Medicine or Physiology in 1983 for her discovery that genes can move around on the chromosomes (transposable elements) – the so-called "jumping genes." She first published the discovery in 1950, but it was not accepted in the scientific community for many years and she worked on her research for decades alone. Her novel idea took 35 years for the Nobel Prize because it was such a revolutionary concept. In addition, the transposable elements that she had conjectured weren't actually seen until the late 1970s when the science of molecular biology had developed significantly farther than it had as of 1950.

McClintock was recognized as one of the brightest geneticists from her graduate school days at Cornell in the 1920s. After serving as an instructor in botany for five years and then working in research for six years, she left Cornell as the university would not appoint women to faculty positions. In the early 1930s, she found chromosomes that formed rings. Later, she found that the ring chromosomes were a

Fig. 5.5 Barbara McClintock. (Courtesy Library of Congress – needs to be cropped)

special case of broken chromosomes. She predicted the existence of structures, which she named telomeres, that would be found on the ends of normal chromosomes. These telomeres maintained a chromosome's stability and integrity but were lost when a chromosome was broken. Telomere research is a rapidly growing area of biology today, with specific implications for cancer and aging. McClintock served as an assistant professor of botany at the University of Missouri for five years. In 1942, she began work at the Cold Spring Harbor Laboratory on Long Island, New York where she would spend the rest of her career.

McClintock was recognized for her genetic work, however, even if the Nobel Prize was slow in coming (the typical span is 10–15 years after the research or discovery). Her name is starred in the seventh edition of *American Men of Science*. She was elected the first woman president of the Genetics Society of America in 1945. In 1944, she was elected to the National Academy of Sciences. In 1970, McClintock was the first woman to receive the National Medal of Science. She also received awards including the Kimber Genetics Award (1967), the Lasker Award (1981), and a lifetime annual tax-free MacArthur genius award starting in 1981. McClintock has been inducted into the National Women's Hall of Fame and was featured on a U.S. postage stamp [13, 15, 17, 18].

5.12   Maria Goeppert-Mayer: Physicist (1906–1972)

The first American woman to receive the Nobel Prize in Physics (in 1963), Maria Goeppert-Mayer (Fig. 5.6) faced discrimination and issues of nepotism throughout most of her professional life. Maria's father, a professor in pediatrics at a German university, encouraged his daughter's intellectual disposition. After finishing the public education available for girls that ended when the girls were 15, she enrolled in a three-year private school that prepared young women for the college examination. She took the entrance examinations one year early, passed (to the teachers' surprise) and entered the University of Göttingen in 1924. Deciding that mathematics was too boring, she became entranced with physics.

Goeppert-Mayer decided to pursue her doctorate in physics and studied for a semester at Cambridge University attending lectures by Nobel Laureate physicist Ernest Rutherford.[4] Upon her return, she was accepted as a doctoral student by physicist and mathematician Max Born who at that time was working with Werner Heisenberg on what would become the Born-Heisenberg theory of quantum mechanics.[5] She married physical chemist Joseph Mayer, an American boarder at the Goeppert house who had completed his postdoctoral work at the University of California, Berkeley and was working at the University of Göttingen as a Rockefeller

---

[4] Sir Ernest Rutherford is considered the father of nuclear physics. He received the Nobel Prize in Chemistry in 1908.

[5] German Max Born would receive the 1954 Nobel Prize in Physics.

Fig. 5.6 Maria Goeppert-Mayer. (Courtesy Library of Congress)

International Education Board Fellow. She then finished her doctorate and the Mayers sailed to Baltimore.[6]

Although she knew she would not have been employed as a woman at a German university, she had hoped to receive better treatment in the U.S. Unfortunately, such was not to be the case. Nepotism rules were very strong and Joseph had been hired as an assistant professor in chemistry at Johns Hopkins University. The physics department had no interest in Goeppert-Mayer both because she was a woman and because her specialty was quantum mechanics. Thus, she decided to do research on a volunteer basis and she was able to convince a reluctant university to give her attic space in the science building. During the ten years that the Mayers were at Johns Hopkins, Goeppert-Mayer at times received a small salary for helping a physics professor with his German correspondence, but was never listed in the catalog. She was given titles like "voluntary assistant" and "research assistant" even though at times she taught some graduate courses and was still performing ground-breaking research.

With the birth of her children Marianne in 1933 and Peter in 1938 and the exodus of many physicists from Germany to the U.S. in the 1930s, Goeppert-Mayer had friends and fellow physicists with whom to discuss her research. In 1940, she and her husband co-authored a textbook, *Statistical Mechanics*, that would become a

---

[6] For her dissertation, she had performed the calculations to support the probability of double photon emissions from electrons orbiting the nucleus of an atom. Thirty years later, laser technology was able to perform the experiments necessary to prove her theory.

classic and go through numerous editions. In 1938, however, Johns Hopkins decided not to grant Joseph tenure, and the Mayers moved to New York City for Joseph to accept a position at twice his salary at Columbia University. Columbia University and its physics department were not interested in having women faculty members either. However, Goeppert-Mayer was able to secure a lectureship in the chemistry department through the efforts of Harold Urey and she taught an interdisciplinary science course at Sarah Lawrence College starting in 1941. At Columbia, she also had the chance to meet and work with Enrico Fermi.

With the start of World War II, the U.S. began to develop an atomic weapon. Goeppert-Mayer worked on the part of the Manhattan Project called SAM – Substitute Alloy Materials project – using her significant scientific knowledge. She published results of her work on uranium isotope separation by photochemical actions and worked with Edward Teller.[7] After the war, the Mayers moved to the University of Chicago, where Joseph was a full professor, but because of strictly enforced nepotism rules, Goeppert-Mayer was an associate professor without salary. She did, however, get an office and get to fully participate in university activities and work with Fermi, Urey, and Teller at the Institute of Nuclear Studies. In addition, Goeppert-Mayer's first graduate student offered her a half-time appointment at Argonne Laboratory, which specialized in nuclear physics. Here she became interested in the occurrence of stable isotopes – the work that would lead to the Nobel Prize. She published her work on spin-orbit coupling and the magic numbers in 1949.

In 1960, the Mayers moved to San Diego and Goeppert-Mayer became a full professor at the University of California – San Diego, after decades of volunteer work in physics. In 1963, she became the second woman – the first was Marie Curie – to win the Nobel Prize in Physics. She shared the prize jointly with Hans Jensen and Eugene Wigner. A headline in the San Diego newspaper of the time exclaimed "S.D. Mother Wins Nobel Prize." She had been elected to the National Academy of Sciences in 1956. Goeppert-Mayer has been inducted into the National Women's Hall of Fame [15, 17–23].

5.13   Grace Murray Hopper: Mathematician and Computer Scientist (1906–1992)

Admiral Grace Murray Hopper (Fig. 5.7) was famous for carrying "nanoseconds" around with her. These lengths of wire – just less than one foot – represented the distance light traveled in a nanosecond, one billionth of a second. She was renowned for trying to convey scientific and engineering terms clearly and coherently to non-technical people.

---

[7] Hungarian-American theoretical physicist Edward Teller is considered the father of the hydrogen bomb.

Fig. 5.7 Admiral Grace
Murray Hopper. (Courtesy
Library of Congress)

Hopper, also known as "Amazing Grace" and "The Grandmother of the Computer Age," helped develop languages for computers and developed the first computer compiler – software that translates English (or any other language) into the zeroes and ones that computers understand (machine language). Actually, her first compiler translated English, French, and German into machine language, but the Navy told her to stick with English because computers didn't understand French and German! Computers truly only understand numbers, but humans can translate those numbers now into English, French, German and even Chinese and Japanese. She was also part of the group that found the first computer "bug" – a moth that had gotten trapped in a relay in the central processor. When the boss asked why they weren't making any numbers, they responded that they were "debugging" the computer. Although Admiral Hopper loved to lay claim to the discovery of this first computer "bug" – and it is in the Smithsonian's National Museum of American History – the term bug had been in use for many years by then.[8]

Hopper received the Society of Women Engineers' Achievement Award in 1964 "in recognition of her significant contributions to the burgeoning computer industry as an engineering manager and originator of automatic programming systems." She was the first woman to attain the rank of Rear Admiral in the U.S. Navy. The Arleigh-Burke-class guided missile destroyer USS Hopper (DDG-70) was commis-

---

[8] Zuckerman reports that Thomas Edison referred to a "bug" in his phonograph as early as 1889. Edison is reported to have defined a bug as "an expression for solving a difficulty, and implying that some imaginary insect has secreted itself inside and is causing all the trouble."

sioned by the U.S. Navy in 1997. Hopper received the National Medal of Technology from President Bush in 1991, the first individual woman to receive the medal: "For her pioneering accomplishments in the development of computer programming languages that simplified computer technology and opened the door to a significantly larger universe of users." She was inducted into the National Women's Hall of Fame in 1994.

Hopper said she believed it was always easier to ask for forgiveness than permission. "If you ask me what accomplishment I'm most proud of, the answer would be all of the young people I've trained over the years; that's more important than writing the first compiler" [24–29].

5.14   Rachel Carson: Marine Biologist (1907–1964)

Best know for her 1962 book *Silent Spring*, that is often credited with starting the 1970s resurgence of the environmental movement in the U.S., Rachel Carson (Fig. 5.8) discovered a love of science while in college. After earning her BS *magna cum laude* from Pennsylvania College for Women (now Chatham College), she spent a summer at the Marine Biological Laboratory at Woods Hole, Massachusetts. An MS in marine zoology at Johns Hopkins University (1932) soon followed.

After a few years of teaching part time at both Johns Hopkins University and the University of Maryland, the death of her father and her sister pushed her to find more lucrative employment to support her mother and two young nieces. She obtained a position as a junior aquatic biologist with the U.S. Bureau of Fisheries in Washington, D.C, after receiving the highest score on a civil service examination. She was one of the first two women to be hired by the department in a nonclerical capacity.

And, she was able to continue pursuing her other love in life besides science – writing. Her first book, *Under the Sea World,* was published in 1941 but did not sell well because of World War II. Her next book, *The Sea Around Us,* was published in 1951 and was a great success, in part because portions had been serialized in the *New Yorker* magazine prior to publication. It was on the *New York Times* best-seller list for 86 weeks. *Under the Sea World* was then re-released and it also became a best seller.

The success of *The Sea Around Us* allowed Carson to write full-time and she resigned from the U.S. Fish and Wildlife Service. In 1955, *The Edge of the Sea* was published. However, it was the 1962 publication of *Silent Spring*, also serialized in the *New Yorker* magazine, that brought her fame and raised environmental awareness in the U.S. It led to the banning of the pesticide DDT and the creation of the Environmental Protection Agency. *Sense of Wonder*, her last book, was published posthumously. In 1980, President Jimmy Carter awarded the Presidential Medal of Freedom to Rachel Carson posthumously. The U.S. Postal Service issued a Rachel Carson stamp in 1981. Carson has been inducted into the National Women's Hall of Fame [13, 15, 30].

Fig. 5.8   Rachel Carson. (Courtesy Library of Congress)

5.15   Ruth Patrick: Limnologist (1907–2013)

Dr. Ruth Patrick (Fig. 5.9) is credited with laying the groundwork for modern water pollution control efforts. Over her sixty-year career, Patrick advanced the field of limnology, which is the study of freshwater biology. She is recognized, along with Rachel Carson, as having ushered in the current concern for the environment and ecology.

Patrick was encouraged by her father, who gave her a microscope when she was seven years old, and told her "Don't cook, don't sew; you can hire people to do that. Read and improve your mind." She studied botany, receiving her undergraduate degree from Coker College (1929) and both her MS (1931) and PhD (1934) degrees at the University of Virginia.

Patrick was originally hired as a "volunteer," without pay – as women scientists at the time were not paid, in 1933 at the Academy of Natural Sciences in Philadelphia, Pennsylvania. Her initial efforts were in microscopy to work with their collection of diatoms, considered to be one of the best collections in the world. Diatoms are microscopic, symmetric single-celled algae with silica cell walls. They are an important part of the food chain of freshwater ecosystems and indicators of water quality. She continued without pay until 1945 while supporting herself through part-time teaching at the Pennsylvania School of Horticulture and making chick embryo slides for Temple University.

She progressed through several positions at the Academy of Natural Sciences and in 1947 became the curator and chairwoman of the Academy's limnology department, which she founded, today called the Patrick Center for Environmental

Fig. 5.9 Ruth Patrick.
(Courtesy of The Academy
of Natural Sciences,
Philadelphia)

Research.[9] From 1973 to 1976, she served as chairwoman of the Academy's board, the first woman to hold that position. Concurrently, she taught at the University of Pennsylvania. Her courses included limnology, pollution biology, and phycology.[10] Her research included taxonomy, ecology, the physiology of diatoms, the biodynamic cycle of rivers, and the diversity of aquatic ecosystems.

Patrick gave a paper in the late 1940s, at a scientific conference on her diatom research. She had matched the types and numbers of diatoms in the water to the type and extent of pollution. This procedure is today used universally but was groundbreaking at the time. To aid in the effort, she invented the diatometer, a clear acrylic device that holds glass microscope slides. The diatometer collects the diatoms from bodies of water: they attach to the slides and grow there. Her research showed that healthier bodies of water contain many species of organisms. The belief that biodiversity – the number and kinds of species – is the key indicator of water health is today known as the Patrick Principle in her honor. The Patrick Principle is the foundation of all current environmental assessments.

Patrick was actively involved in the drafting of the federal Clean Water Act, passed in 1972. She was called the foremost authority on America's river systems. Patrick estimated at one point that she had waded into 850 different rivers around the globe including the Amazon River.

---

[9] In 2011, the Academy of Natural Sciences became affiliated with Drexel University.

[10] Phycology is the study of algae.

Patrick was the first woman to serve on the board of directors of the DuPont Corporation and was its first environmental activist. She also advised Presidents Johnson and Reagan as well as governors on water quality issues. She served on water pollution and water quality panels for the National Academy of Sciences and the U.S. Department of Interior as well as other federal advisory groups.

Patrick was elected to the National Academy of Sciences in 1970, as the twelfth woman to receive this form of recognition. She received the National Medal of Science in 1996, was elected a Fellow of the American Academy of Arts and Science, was inducted into the National Women's Hall of Fame, and was the recipient of over 25 honorary degrees [13, 15, 16, 31–35].

5.16   Rita Levi-Montalcini: Neurobiologist (1909–2012)

Rita Levi-Montalcini's love of research survived Italy's fascist government's World War II edicts that forbade Jews from practicing medicine or science. She ground her sewing needles into the implements that she needed – scalpels and spatulas – and used ophthalmologist miniature scissors and watchmakers forceps to perform surgery on chicken embryos. She had decided to examine the development of the nervous system. How could she focus on solving a neuroembryological problem with German armies spreading death and destruction throughout Europe? Many years later, she said, "The answer lies in the desperate and partially unconscious desire of human beings to ignore what is happening in situations where full awareness might lead one to self-destruction." Because Italian journals would not accept the papers she wrote on her research, the papers instead were published by Belgian and Swiss publications that could fortuitously be read by researchers in the U.S.

And that is how Levi-Montalcini found herself on a ship to the U.S. in 1946 planning to spend a few months at Washington University in St. Louis. She spent almost three decades. Levi-Montalcini's research led her to the discovery of the nerve growth factor (NGF) in 1952 for which she received the Nobel Prize in Physiology or Medicine in 1986 jointly with her colleague Stanley Cohen. Growth factors have many current and potential applications in medicine including speeding up the rate of burn healing and diminishing the side effects of chemotherapy and radiation treatments. Her discovery of NGF also transformed the way that the nervous system is viewed.

Homesick for Italy, Levi-Montalcini established the Institute of Cell Biology of the Italian Research Council in Rome and began to spend six months each year in the U.S. and six months in Italy. That continued until 1977, when she returned to Italy full time.

In addition to the Nobel Prize, Levi-Montalcini was elected to the National Academy of Sciences in 1968 (the tenth woman) and was the first woman to become a member of the Pontifical Academy of Sciences in Rome (1974). Among her many other awards, she received the U.S. National Medal of Science and a number of honorary degrees [15, 17, 18].

5.17   Dorothy Crowfoot Hodgkin: Chemist/Crystallographer (1910–1994)

Nobel Laureate Dorothy Crowfoot Hodgkin loved to take on biochemical problems that were deemed too difficult to solve. Thus, in 1964 when she received the Nobel Prize in Chemistry (the third woman to receive the honor and the only British woman), it was fitting that the Nobel Committee specifically mentioned her work to identify the structure of penicillin and vitamin B-12, two "unsolvable" molecules. That she did this work with substandard equipment and hands crippled with rheumatoid arthritis just further demonstrates her unwavering commitment to her science.

Hogkin discovered chemistry and crystals as a child. When she decided that she wanted to attend Oxford University to study chemistry, she learned that she was not prepared for their entrance examination. But she was never one to let any obstacle stand in her way. With tutoring and study, she passed the examination. After her graduation, Hodgkin was finally able to find a job in a laboratory at Cambridge University where she worked with x-ray crystallography. She was then enticed to return to Oxford where she would spend the rest of her career.

During World War II, Hodgkin and her graduate student Barbara Low were able to determine the structure of penicillin. As a result of her reputation, in 1948, she was approached about working out the structure of vitamin B-12. She was able to announce the structure in 1956, eight years after she began her work. In 1969, she worked out the structure of insulin.

Hodgkin received many other awards and honors in addition to the Nobel Prize. In 1965, when she received a letter from Buckingham Palace, she didn't even want to open it. She was happy when she eventually did, however. It informed her that Queen Elizabeth II would be giving her the Order of Merit – making her only the second woman to have received it – the first being Florence Nightingale. Hodgkin was a Fellow of the Royal Society (the third woman elected to the organization in 1947) and was elected to membership in the American Academy of Arts and Sciences. An asteroid is named for her [15, 18, 36].

5.18   Chien-Shiung Wu: Physicist (1912–1997)

One of the world's foremost nuclear physicists, Chien-Shiung Wu (Fig. 5.10) is best known for a classic experiment in beta decay that confirmed a theory put forward a year earlier by two physicists, Tsung-Dao Lee and Chen Ning Yang, for which Lee and Yang received the 1957 Nobel Prize in Physics. The theory was that parity was not always conserved in nature. For beta particles, the prediction was that the result of beta decay would not be symmetrical because of certain weak forces. Wu's experiments showed clear evidence that violation of parity conservation was observed and the results were announced in 1957.

Wu was born near Shanghai, China and was encouraged to pursue an education by her revolutionary father. She entered the National Central University intent on

Fig. 5.10 Chien-Shiung Wu. (Courtesy Library of Congress)

studying physics and received her degree in 1934. She left China to pursue graduate training and ended up at the University of California at Berkeley, influenced by the presence of Ernest Orlando Lawrence, inventor of the atom-smashing cyclotron and by her husband-to-be physicist Chia-Liu Yuan. Wu received her Ph.D. in 1940, staying at Berkeley for two years, Smith College for a year, and Princeton University for a short period (she was the first woman instructor ever hired at Princeton), before she began work on the Manhattan Project through a position at Columbia University. Wu eventually became a full professor at Columbia and retired from Columbia in 1981.

Although she did not receive a Nobel Prize for her efforts, Wu was elected to the National Academy of Sciences in 1958. She received the first Wolf Prize awarded by Israel (1978), the first Research Corporation Prize given to a woman (1959), the Comstock Award of the National Academy of Sciences (1954), and the National Medal of Science in 1975. She was the first living scientist with an asteroid named after her (1990). Posthumously, Wu was inducted into the National Women's Hall of Fame [15, 17, 19].

5.19 Mary Leakey: Paleoanthropologist (1913–1996)

Introduced to cave paintings and prehistoric flint tools as a child, Mary Leakey learned that archeology was a possible career for women after meeting Dorothy Liddell, a pioneering British archeologist. Leakey took classes in geology and archeology at the University of London and obtained field experience by participating in excavations. Her focus on the excavations in which she participated early on were stone tools.

At her first meeting with archaeologist Louis Leakey, who would later become her husband, she agreed to assist him with illustrations for his 1934 book, *Adam's Ancestors: An Up-to-Date Outline of What Is Known about The Origin of Man*. Their excavation together in 1934 in Clacton, England led to her first publication which was a report printed in 1937 in the *Proceedings of the Prehistoric Society*.

She accompanied Louis Leakey to the Olduvai Gorge in the East African country of Tanzania for the first time in 1935. During a 1959 excavation rhere, she found fragments of a human skull. Later she would participate in excavations in Kenya. Her work at another site in Tanzania, Laetoli, led to her discovery of two trails of hominid footprints that were dated to 3.5 million years ago. This discovery moved the date for the development of upright postures for hominids to much earlier than had been previously believed. She considered this her most important discovery. Deemed one of the most important paleoanthropologists of the twentieth century, Leakey was always more concerned with the discovery itself and left to others determining the meaning of those discoveries [15].

5.20   Anna Jacobson Schwartz: Economist (1915–2012)

Called "one of the world's great monetary scholars," Anna Jacobson Schwartz (Fig. 5.11) became interested in economics while she was in high school. She earned her undergraduate degree at Barnard College, and her MA in economics (1935) and PhD (1964) from Columbia University. She spent more than seventy years at the National Bureau of Economic Research starting in 1941.

Her initial publications were in the field of quantitative economic history. Working with economist and Nobel Laureate Milton Friedman, she produced three volumes on American and British monetary history leading to the sobriquet "the high priestess of monetarism." One of those volumes, *A Monetary History of the United States, 1867–1960*, continues to be one of the most widely cited texts today. It concludes that the Federal Reserve was a significant cause of the Great Depression. Her efforts demonstrated the importance of the money supply and the need for that supply to be stable and predictable.

Her American Economics Association Distinguished Fellow citation (1993) said that she was "a leading authority on economic history, monetary economics, international monetary systems, and monetary statistics." Her many other honors included Fellow of the American Academy of Arts and Sciences, multiple honorary degrees and induction into the National Women's Hall of Fame [37–39].

Fig. 5.11 Anna Jacobson Schwartz. (Courtesy of David Shankbone/Wikipedia)

5.21 Gertrude Belle Elion: Biochemist (1918–1999)

Gertrude Elion carried out drug research that has saved the lives of thousands of people despite discrimination against her in the workplace and in academia due to her gender. Elion received the Nobel Prize in Physiology or Medicine in 1988 along with George Hitchings and Sir James W. Black for her work in drug research. Their research had demonstrated the differences in nucleic acid metabolism between normal cells and disease-causing cancer cells, protozoa, bacteria, and viruses.

Elion grew up in New York City and graduated from Hunter College with highest honors in chemistry (Phi Beta Kappa) in 1937. She was turned down for graduate assistantships from fifteen schools because of her gender. Discouraged from seeking a position in a laboratory because she was told that as a woman she might district the male chemists, Elion took secretarial classes for a short time. By volunteering at a chemistry lab and eventually being paid a small salary, she earned enough money to pay for graduate school for one year. She had a variety of positions including teaching and quality control, before she called the Burroughs Wellcome Company to find out if they had a research laboratory and, after interviewing on a Saturday, went to work for them in 1944. She would stay until her retirement in 1983.

At Burroughs Wellcome, Elion had a chance to work in the areas of organic chemistry, biochemistry, pharmacology, immunology, and virology. She also made significant drug discoveries. Her childhood leukemia drug, 6-MP, was approved for treatment and sold under the name of Purinethol. She then developed thioguanine, in 1950, also used for treatment of children with leukemia. Today, thioguanine is used primarily to treat acute myelocytic leukemia in adults. Another of Elion's drugs, 57-322 (which later became known as azathioprine and marketed as Imuran), allow organs to be transplanted into people as it suppresses immune system rejections. The drug is also used to treat autoimmune lupus, anemias, hepatitis, and severe rheumatoid arthritis.

Her compound allupurinol saves people with kidney blockage. It is also used to treat Leshmaniasis disease, a major problem in South America. In 1968, Elion revisited one of her earlier drug failures and turned it into one of her biggest successes. She developed Acyclovir which is used to treat shingles, Epstein-Barr virus, pseudorabies, and herpes encephalitis. It is marketed as the drug Zovirax. The approach Elion developed was used at Burroughs Wellcome to develop the AIDS drug Azidothymidine, called AZT.

Elion was elected to the National Academy of Sciences in 1991 and received the National Medal of Science that same year. Burroughs Wellcome gave Hitchings and Elion $250,000 apiece to donate to a charity of their choice in recognition of their contributions to science. Elion gave the money to Hunter College for scholarships for women in chemistry and biochemistry. In 1991, Gertrude Belle Elion was the first woman elected to the National Inventors Hall of Fame. She has been inducted into the National Women's Hall of Fame [15, 18, 19].

5.22   Katherine Johnson: Mathematician (1918–1920)

Demonstrating her significant intelligence early in her life, Katherine Johnson (Fig. 5.12) was attending high school at age 13 on the campus of the historically black West Virginia State College. She graduated from that same institution with highest honors in 1937 and began to teach in the West Virginia public schools. When West Virginia quietly integrated its graduate schools in 1939, Johnson and two men were hand-picked to integrate West Virginia University. She left graduate school without getting her degree for marriage and a family and returned to teaching when her children got older.

When Johnson learned about opportunities at the National Advisory Committee for Aeronautics (NACA), the predecessor to what is today the National Aeronautics and Space Administration (NASA), she and her family moved to Virginia and she began work at the Langley Research Center in Hampton, Virginia in 1953. After the launch of Sputnik in 1957, the course of her career changed as it did for all of the Langley Research Center employees: the focus was now beating the Russians in space.

Johnson calculated the trajectory analysis for Astronaut Alan Shepard's Freedom 7 mission in 1961. Her name appears on a 1960 NASA report, the first time that a woman's name was credited on such a document. As preparations were ongoing for Astronaut John Glenn's Freedom 7 flight in 1962, Johnson was called upon to per-

Fig. 5.12 Katherine
Johnson. (Courtesy NASA)

form the same calculations as the new-fangled computers were doing for the complex orbital flight. In fact, John Glenn insisted that she perform those calculations; he had confidence in her abilities and was skeptical of those machines. This interaction is famously portrayed, although sensationalized, in the 2016 movie *Hidden Figures,* which brought Johnson and the other African-American women computers at NASA out of the shadows. In 2015, at age 97, Johnson received the Presidential Medal of Freedom, the highest civilian honor awarded in the U.S., from President Obama [40].

5.23   Julia Robinson: Mathematician (1919–1985)

A trailblazing pioneer for women, Julia Robinson was the first female mathematician to be elected to the National Academy of Sciences (1975), the first female president of the American Mathematical Society, and the first woman mathematician to receive a MacArthur Fellowship Foundation. Although she preferred to be remembered for her contributions to mathematics, Julia Robinson's personal story also needs to be told.

Robinson overcame a variety of tragedies in her youth. Her mother died when she was two years old. Robinson was afflicted with rheumatic fever after having scarlet fever and missed two years of school. But she did catch up and received honors in mathematics and science when she graduated from high school as well as the Bausch-Lomb medal for all-around excellence in science. Robinson enrolled at San Diego State University and was able to continue her education even after her father committed suicide. She transferred to and graduated from the University of California at Berkeley, where she would later earn her PhD and meet her husband, Raphael Robinson.

Robinson worked for twenty-years on the tenth problem on Hilbert's list and provided a significant foundation for Yuri Matijasevic's solution.[11] The problem is to find an effective method for determining if a given Diophantine equation is solvable for integers (Diophantes lived in the latter part of the third century). She also worked at RAND Corporation where she made what is considered a fundamental contribution to game theory.

Upon her election to the National Academy of Sciences, she was made a full professor at the University of California, Berkeley. Always interested in being seen for her ability and not her gender, she debated with herself as to whether or not she would accept the presidency of the American Mathematical Society:

In 1982 I was nominated for the presidency of the American Mathematical Society. I realized that I had been chosen because I was a woman and because I had the seal of approval, as it were, of the National Academy. After discussion with Raphael, who thought I should decline and save my energy for mathematics, and other members of my family, who differed with him, I decided that as a woman and a mathematician I had no alternative but to accept. I have always tried to do everything I could to encourage talented women to become research mathematicians. I found my service as president of the Society taxing but very, very satisfying.

Robinson was also elected to the American Academy of Arts and Sciences (1985), received an honorary degree and was the second woman to deliver the Colloquium Lectures of the American Mathematical Society [15, 16, 41–43].

5.24   Rosalind Franklin: Biologist (1920–1958)

Rosalind Franklin made key contributions to the structures of coals and viruses and provided the scientific evidence about the double-helix structure of DNA for which American biologist James Watson, British physicist Francis Crick, and British physicist and molecular biologist Maurice Wilkins shared the Nobel Prize in 1962. Although Nobel Prizes are only awarded to living scientists, her contributions to the effort to discover the structure of DNA are thought by some to have been overlooked.

Franklin grew up in London and decided at any early age to pursue a career in science. She graduated from Cambridge in 1941 and after a short-lived research scholarship to study gas-phase chromatography with future Nobel laureate Ronald

---

[11] In 1900, German mathematician David Hilbert published 23 problems in mathematics that were unsolved at the time.

G.W. Norrish, accepted a job as assistant research officer with the British Coal Utilization Research Association (CURA). At the CURA, she applied her knowledge of physical chemistry to study the microstructures of coal. In 1947, she moved to Paris where she learned the technique known as x-ray diffraction. In 1951, she left Paris to set up an x-ray diffraction unit in a laboratory at St. John T. Randall's Medical Research Council at Kings' College in London to produce diffraction pictures of DNA.

Here she worked with Maurice Wilkins, who took an intense dislike to her. Wilkins would later show Watson the DNA diffraction pictures that Franklin had amassed (without her permission) and here Watson saw the evidence needed to discern the helical structure of DNA. Franklin had recorded in her laboratory notebook that DNA had a helical structure of two chains prior to the publication by Watson and Crick of their similar analysis.

Franklin left King's College for Birkbeck College where she worked on the tobacco mosaic virus, particularly the RNA structure and the location of protein units. She died at age 37 from ovarian cancer [13].

5.25   Rosalyn Yalow: Physicist (1921–2011)

Nobel Laureate Rosalyn Yalow (Fig. 5.13) was the first American-educated woman to receive a Nobel prize in the sciences. She shared the 1977 Nobel Prize in Physiology or Medicine for her work on the development or radioimmunoassay

Fig. 5.13 Rosalyn Yalow. (Courtesy Library of Congress)

(RIA) on peptide hormones. RIA is a technique used to screen for very tiny amounts of chemicals in biological tissues and fluids and has revolutionized the treatment of hormonal disorders including diabetes.

Rosalyn Yalow grew up in New York City and entered Hunter College at age 15 having already selected physics as her major and intending to go to medical school. Upon graduation, however, she took a job as a secretary as American medical schools were not admitting Jewish men, much less Jewish women. However, she was fortunate to receive a fellowship in physics from the University of Illinois because World War II was imminent and male graduate students were being lost to the draft. She was the first woman in the University of Illinois engineering school since 1917, when World War I had been underway.

On her first day of graduate school, Yalow met her future husband, Aaron Yalow. They did PhD research in nuclear physics together and were married in 1943. In 1945, Yalow received her PhD, the second woman to receive a PhD in physics at the University of Illinois. By 1947, she had joined the Bronx VA hospital in order to set up a radioisotope service although she was teaching at Hunter College. In 1950, she resigned from Hunter College and began to look for a collaborator. In physician Solomon Berson, she found the partner to focus on the medical aspects of her research while she focused on the engineering and physics aspects. They collaborated until his death in 1972.

RIA was developed as a side issue to insulin research by Berson and Yalow. When adult diabetics were injected with radioactively tagged insulin, the insulin was retained longer than normal controls indicating that the people who had taken insulin had developed antibodies to it. Unfortunately, this conclusion flew in the face of conventional wisdom and their initial conclusions were met with some skepticism. However, they persevered and developed RIA in a wide variety of areas that made endocrinology one of the hottest areas of medicine. Today, RIA helps scientists and medical professionals measure the concentrations of hormones, vitamins, viruses, enzymes, and drugs, among other substances. Berson and Yalow decided not to patent RIA but instead make it available to the medical community.

Understanding her role as a trailblazer for women scientists, Yalow said:

> We cannot expect that in the foreseeable future women will achieve status in academic medicine in proportion to their numbers. But if we are to start working towards that goal we must believe in ourselves or no one else will believe in us; we must match our aspirations with the guts and determination to succeed; and for those of us who have had the good fortune to move upward, we must feel a personal responsibility to serve as role models and advisors to ease the path for those who come afterwards.

Yalow was elected a member of the National Academy of Sciences in 1975. She was the first woman to win the Albert Lasker Basic Medical Research Award in 1976, which is often a forerunner to the Nobel Prize. In 1988, she received the National Medal of Science. Yalow has been inducted into the National Women's Hall of Fame [13].

References

1. M.W. Rossiter, *Women Scientists in America: Before Affirmative Action 1940–1972* (The Johns Hopkins University Press, Baltimore, 1995)
2. A.M. Barker, Women in Engineering During World War II: A Taste of Victory, 21 Nov 1994, unpublished, Rochester Institute of Technology
3. U.S. Department of Labor, Women's Bureau, *Employment Opportunities for Women in Professional Engineering*, Women's Bureau Bulletin No. 254, Washington, DC: U.S. Government Printing Office, 1954
4. B. Harris, *Beyond her Sphere: Women in the Professions in American History* (Greenwood Press, Connecticut, 1978)
5. W.K. LeBold, D.J. LeBold, Women Engineers: A Historical Perspective. ASEE Prism **7**, 30–32 (1998)
6. U.S. Department of Labor, Women's Bureau, *Careers for Women in the Physical Sciences*, Women's Bureau Bulletin 270, 1959
7. J.C. Lucena, 'Women in Engineering' a history and politics of a struggle in the making of a statistical category, in *Proceedings of the 1999 International Symposium on Technology and Society – Women and Technology: Historical, Societal, and Professional Perspectives*, pp. 185–194, New Brunswick, 29–31 July 1999
8. P.J. Read, B.L. Witlieb, *The Book of Women's Firsts* (Random House, New York, 1992)
9. E.S. Green, *United States House of Representatives, History, Art & Archives*, https://history.house.gov/People/Detail/14080. Accessed 11 May 2020
10. J.A. Baer, *Women in American Law: The Struggle Toward Equality From the New Deal to the Present*, 2nd edn. (Homes & Meier, New York, 1996)
11. S. Tobias, *Faces of Feminism: An Activist's Reflections on the Women's Movement* (Westview Press, Boulder, 1997)
12. E. Babco, *Professional Women & Minorities: A Total Human Resources Data Compendium*, Commission on Professionals in Science & Technology, 13th edn. Apr 2000. ISSN: 0190-1796
13. B. F. Shearer, B. S. Shearer (eds.), *Notable Women in the Life Sciences* (Greenwood Press, Westport, 1996)
14. R.S. Baldwin, *The Fungus Fighters: Two women scientists and their discsovery* (Cornell University Press, New York, 1981)
15. P. Proffitt (ed.), *Notable Women Scientists* (The Gale Group, Detroit, 1999)
16. M.J. Bailey, *American Women in Science: A Biographical Dictionary* (ABC-CLIO, Denver, 1994)
17. S.B. McGrayne, *Nobel Prize Women in Science: Their Lives, Struggles, and Momentous Discoveries* (Carol Publishing Group, New York, 1993)
18. B. F. Shearer, B. S. Shearer (eds.), *Notable Women in the Physical Sciences* (Greenwood Press, Westport, 1997)
19. E. Rutherford. https://en.wikipedia.org/wiki/Ernest_Rutherford. Accessed 11 May 2020
20. M. Born. https://en.wikipedia.org/wiki/Max_Born. Accessed 11 May 2020
21. Mayer Papers, 1920–1983 (MSS 0047), https://library.ucsd.edu/speccoll/findingaids/mss0047.html. Accessed 11 May 2020
22. E. Teller. https://en.wikipedia.org/wiki/Edward_Teller. Accessed 11 May 2020
23. Grace Hopper 1906–1992. www.greatwomen.org/hopper.htm. Accessed 1 Sept 1999
24. www.swe.org/SWE/Awards achieve3.htm. Accessed 1 Sept 1999
25. Grace Hopper, National Medals of Science and Technology Foundation. https://www.nationalmedals.org/laureates/grace-hopper. Accessed 9 Apr 2020
26. C.W. Billings, *Grace Hopper: Navy Admiral and Computer Pioneer* (Enslow Publishers, Inc., Hillside, 1989)
27. L. Zuckerman, Think tank: If There's a bug in the etymology, you may never get it out. *The New York Times*, 22 Apr 2000

28. A. Stanley, *Mothers and Daughters of Invention: Notes for a Revised History of Technology* (Rutgers University Press, New Brunswick, 1995)
29. G. Kass-Simon, P. Farnes (eds.), *Women of Science: Righting the Record* (Indian University Press, Bloomington, 1990)
30. R. Patrick, A Pioneer in Science and Pollution Control Efforts, Is Dead at 105. http://www.nytimes.com/2013/09/24/us/ruth-patrick-a-pioneer-in-pollution-control-dies-at-105.html. Accessed 16 Apr 2020
31. T.L. Bott, B.W. Sweeney, A Biographical Memoir, Ruth Patrick 1907–2013, National Academy of Sciences (2014). http://www.nasonline.org/publications/biographical-memoirs/memoir-pdfs/patrick-ruth.pdf
32. The Academy of Natural Sciences of Drexel University. http://www.ansp.org/about/drexel-affiliation/. Accessed 16 Apr 2020
33. R. Patrick. https://www.nationalmedals.org/laureates/ruth-patrick. Accessed 16 Apr 2020
34. The Academy of Natural Sciences of Drexel University, Dr. Ruth Patrick. https://ansp.org/research/environmental-research/people/patrick/. Accessed 16 Apr 2020
35. M. Ogilvie, J. Harvey (eds.), *The Biographical Dictionary of Women in Science: Pioneering Lives from Ancient Times to the Mid-20th Century* (Routledge, New York, 2000)
36. B. Shiels, *Winners: Women and the Nobel Prize* (Dillon Press, Inc., Minneapolis, 1985)
37. A.J. Schwartz. https://jwa.org/encyclopedia/article/schwartz-anna-jacobson. Accessed 12 May 2020
38. J.S. Anna. https://www.womenofthehall.org/inductee/anna-jacobson-schwartz/. Accessed 12 May 2020
39. A. Schwartz. https://www.womenofthehall.org/inductee/anna-jacobson-schwartz/. Accessed 12 May 2020
40. "Katherine Johnson Biography", by Margot Lee Shetterly, https://www.nasa.gov/content/katherine-johnson-biography. Accessed 16 Apr 2020
41. C. Morrow, T. Perl (eds.), *Notable Women in Mathematics: A Biographical Dictionary* (Greenwood Press, Westport, 1998)
42. J. Robinson. https://en.wikipedia.org/wiki/Julia_Robinson. Accessed 10 Apr 2020
43. Hilbert's problems. https://en.wikipedia.org/wiki/Hilbert%27s_problems. Accessed 11 May 2020

Chapter 6
Science in the New Millenium

6.1 Introduction

During the last quarter of the twentieth century and into the twenty-first century, many efforts were undertaken to level the playing field for women in science. Margaret Rossiter, the renown history of science author, relates in the third volume of the history of women scientists in America [1]:

> The overall tale in the years 1970-2010 is one of a small number (50-100) of marginal individuals finding one another, forming organizations and electing leaders, starting and running grass-roots campaigns, and raising funds for projects that, while not centrally directed, collectively made inroads into a sexist and elitist system over several decades... There were many legal battles and numerous firsts at each step, for everything needed to be changed at nearly every institution... Individuals created many new events, traditions, awards, that increase the positive feedback and cultural atmosphere that a woman entering science or engineering might encounter...

6.2 Scientific Organizations

The heady years following the resurgence of the women's movement and increasing numbers of women in colleges and the workforce pursuing scientific careers resulted in an increasing number of women's scientific organizations and women's committees or women's caucuses of existing scientific organizations. The Association for Women in Psychology was formed in August 1969. The women's committee of the American Anthropological Association was established in February 1970. In April 1971, a women's committee of the American Physical Society was convened. The Association for Women in Mathematics was established in January 1971. In April 1971, the Association for Women in Science (AWIS) was formalized. The Association for Women in Computing was founded in 1978 [1–3].

J. S. Tietjen, *Scientific Women*, Women in Engineering and Science,
https://doi.org/10.1007/978-3-030-51445-7_6

The Chemists' Club of New York City, which had been founded in 1921, admitted women in 1971. The Biosystematists of the Bay Area, whose meetings were held at the Men's Faculty Club at the University of California, Berkley, admitted women in 1971. The Nuttall Ornithological Club, based in the Boston area and established in 1873, admitted women in the mid-1970s. Others held out longer. The Explorers Club of New York City voted to admit women in April 1981. Two days before the U.S. Supreme Court upheld a New York City human-rights law banning sex bias in membership in certain private clubs, the heretofore all-male Cosmos Club, where scientists in the New York City area gathered, voted to admit women. That was in June 1988 [1].

Yet, for many of these organizations, a way of clinging to the male past was through the "Fellow" recognition, bestowed as an elected honorary status. The societies were slow to elect women as fellows and many older women scientists saw the younger ones honored but not themselves. In 1984, women constituted 32.8% of the membership of the American Psychological Association yet only 16.2% of its fellows. By 1988, women constituted 37.1% of the membership yet the percent of women fellows reached only 17.7%. As another example, in 1994, women constituted 15% of the membership of The American Geophysical Union, but only 3.1% of its fellows [1].

In 1972, the American Association for the Advancement of Science (AAAS) established the Office of Opportunities in Science (OOS) to advocate in Washington, DC for the concerns of women and minority scientists. When it came time for the AAAS to nominate someone to the National Science Board (NSB) which oversees the National Science Foundation (NSF), the director of OOS nominated biologist and academic administrator Jewel Plummer Cobb. In 1974, Cobb became the first black woman and one of the youngest people to serve on the NSB. Among its many activities, OOS was responsible for organizing a weekend meeting of minority women scientists in 1975 out of which came a groundbreaking AAAS report *The Double Bind: The Price of Being a Minority Woman in Science*. Later, disabled people would be added to the OOS portfolio and the OOS would publish the best-selling *Barrier Free Meetings* [1].

More than seventy women's organizations, women's committees of existing organizations or women's caucuses were in existence by 1978. In addition, an umbrella organization of a sort was established – the Federation of Organizations for Professional Women (FOPW). The intent of this organization was to lobby Congress on behalf of the organizations who were its constituents. FOPW stayed in existence until the early 1990s [1].

Something else new happened in the scientific organizations in the 1970s as a result of the women's organizations, committees and caucuses. Women were voted into officerships, including as President. When they finished their presidential terms, those women were now distinguished elders of the organization. The American honorary academics increased their numbers of women members significantly and the number of women receiving the Nobel Prize in the sciences could no longer be counted on one's fingers. And in 1984, the National Academy of Sciences council voted to change all male pronouns in its constitution to gender-neutral ones! [1].

6.3   1972 Legislation

The Executive Orders of the 1960s and voluntary actions on the parts of institutions had not moved the needle for women in science. Bernice "Bunny" Sandler is a prime example. Clinical psychologist Bunny Sandler filed a protest with the Department of Labor under Executive Order 11246, after having been denied several jobs at the University of Maryland because "she came on too strong for a woman." Eventually, Sandler would file complaints against 260 universities. As she spoke at many campuses over the years, she advised the women scientists to take action. For a while, the Equal Employment Opportunity Commission (EEOC) under Eleanor Holmes Norton did provide a glimmer of hope but voluntary actions by universities did not yield much progress. Fortunately, 1972 would prove to be a pivotal legislative year [1].

Two landmark pieces of legislation were enacted in 1972 that would help advance the cause for equality for women in science. These were the 1972 passage of the Education Amendments Act (particularly Title IX) and the Equal Employment Opportunity Act of 1972. Bunny Sandler worked with Representative Edith Green (Oregon) and others on what became Title IX of the Education Amendments Act of 1972. The acts put teeth into affirmative action. With their enactment, organizations and individual women now had the basis on which to sue for equal treatment [1].

6.3.1   Education Amendments Act

Title IX of the Education Amendments Act prohibited discrimination on the basis of sex in all federally-assisted educational programs. Title IX stated in part, "No person in the United States shall, on the basis of sex, be excluded from participation in, be denied the benefits of or be subjected to discrimination under any education program or activity receiving federal financial assistance."

Title IX extended the Equal Pay Act of 1963 and Title VII of the Civil Right Act of 1964 to educational workers and applied to admissions of females to all public undergraduate institutions, professional schools, graduate schools, and vocational schools. A very significant consequence of this act was that caps on the numbers of women students accepted into medical, law, business, and other professional schools were finally abolished [4].

6.3.2   Equal Employment Opportunity Act of 1972

The Equal Employment Opportunity Act of 1972 marked a significant enhancement of the actions that could be taken by the EEOC and an expansion of the organizations to which the law applied. The most significant of these as far as women in science were concerned included [5]:

- The EEOC could now sue nongovernment respondents – employers, unions, and employment agencies – on behalf of individuals who had experienced discrimination.
- Educational institutions were now subject to Title VII of the Civil Rights Act. Congress noted that discrimination against women and minorities was just as pervasive in educational institutions as it was in any other area of employment.
- The size of the organization that was subject to the Act was reduced from 25 to 15 employees.

6.4   Lawsuits

The legislation enacted in 1972 said the law of the land was equal pay and equal opportunity. Women and organizations for whom equal pay and equal opportunity were not the order of the day filed lawsuits to advance the standing of, funding for and equality of themselves and women scientists in general.

An important and groundbreaking lawsuit was filed by AWIS. Many of the women who helped found AWIS were biologists who were frustrated with the paucity of women on the National Institute of Health's (NIH) technical panels and study sections and, correspondingly, the few grants that were awarded to the women who applied. After an AWIS delegation meeting with the NIH and the Department of Health, Education and Welfare (HEW) in late 1971 that had an unsatisfying outcome, AWIS did the heretofore unthinkable – in March 1972, they filed suit against HEW (which oversaw the NIH). The lawsuit was joined by other organizations including the National Organization for Women, the Association for Women in Mathematics, and the Association for Women in Psychology. Only months after the lawsuit filed, the percent of women on the NIH study sections increased from 2% to 20%. Against all odds, in 1973, the judge ruled in favor of AWIS [1].

Three pioneering academic discrimination cases in the 1970s set precedents for women in academia and the universities themselves around tenure. These suits were enabled by the Equal Employment Opportunity Act of 1972 that amended Title VII of the Civil Rights Act. Through this legislation, individuals after receiving a letter from the EEOC, could sue to protest a denial of employment or tenure or for equal pay. Other cases were fought on the topic of equal pay.

Biochemist Sharon L. Johnson filed suit against the University of Pittsburgh Medical School in 1973. She had been hired in 1967 and was coming up for tenure, which she had been assured when she was hired would not be a problem. Her tenure was denied, there was no appeal process, and she was to leave within eighteen months. In May of 1973, a judge awarded her an injunction against the university so that it was forced to keep her until her NIH grants ran out. The university's attorneys had never experienced this before and it attracted a lot of press. Eventually the stalling tactics and superior financial position of the university and the judge's refusal to rule on her case even after 74 days of courtroom procedures and four years took their toll. Johnson decided not to appeal his refusal to award her tenure or to rule on the merits of the case [1].

Despite Johnson's loss, there was a heartening development in 1977. In September 1977, a month after Johnson opted not to appeal, anthropologist Louise Lamphere settled her groundbreaking class action suit against Brown University. Lamphere, like Johnson, had been hired and then denied tenure. The President of Brown University, wishing to avoid negative publicity and proceed with a major fundraising drive, signed a consent decree without going to court. Under the decree, the University would be monitored for at least a decade to ensure that it added women to its faculty. In addition, Lamphere as well as two other women would be awarded tenure retroactively and a fourth woman would receive a monetary settlement. Brown University surprisingly agreed to pay the women's legal fees. Other women scientists saw the Brown-Lamphere settlement decree as a means of avoiding a costly, lengthy court battle [1].

The 1980 decision of a case filed in 1973 by Shyamala Rajender against the University of Minnesota encouraged women scientists around the U.S. Rajender had a series of one-year positions as research associate and instructor but was not being interviewed for tenure-track positions. She eventually hired a high-powered attorney who took her case on a contingency basis. He made the lawsuit a class action to include all women discriminated against by the University of Minnesota system since 1972. The out-of-court settlement in 1980 was in Rajender's favor, required a monitoring system of future hires in the university system and required the university to pay Rajender's legal fees [1].

These lawsuits and the attendant publicity and costs led to changes at universities. Positions began to be advertised. Search committees were established. Procedures were followed and decisions were documented. Faculty senates adopted more grievance procedures when tenure was denied. These actions allowed women scientists something that had long been denied most of them – a tenured position at a research university [1].

Sex discrimination lawsuits were also filed against corporations, non-profits, and the federal government some of whom also appeared to be dragging their feet on these issues. One very significant case was filed by the EEOC with the Federal Communications Commission in December 1970. This case argued that AT&T should not be granted its requested rate increase due to its violations of state and federal prohibitions against discrimination in employment. At that time, AT&T's employees were generating about seven percent of the EEOC's total complaints. This was a novel and untested avenue for discrimination actions. It was successful. Considered a landmark decision, the consent decree was signed in January 1973. AT&T, then the country's largest employer, agreed to make $38 million in restitution through payments to women and minority employees and the implementation of immediate pay increases to women and minority employees [1, 6, 7].

Judith Osmer, a chemist at The Aerospace Corporation filed the first class-action lawsuit involving technical women in 1971. She alleged sex discrimination. The case was in the courts for 12 years but the result was that The Aerospace Corporation was found guilty on all counts by all three agencies with which it was filed – the California Fair Employment Practices Commission, the EEOC and the Contract

Compliance Office of the Department of Defense. The lawsuit is credited with changing employment practices at all U.S. defense contractors [1, 8].

Mathematician Rosalind Marimont filed a class-action lawsuit in 1973 on behalf of 7000 women at the NIH for denial of promotions. When the case was finally settled in 1979, Marimon was awarded back pay and attorney's fees. In addition, the settlement required the NIH to develop "fair and definite" promotion procedures. However, disparities in grade levels and associated pay between men and women persisted for decades [1, 9].

Also in 1973, forester Gene Bernardi at the U.S. Forest Service (USFS) filed a class-action lawsuit alleging sex discrimination in hiring and promotions. There were very few women in the low management ranks and none in the top ranks at the USFS. The case never went to trial. After much discovery, in 1981, a judge approved a consent decree awarding money to Bernardi and compliance reporting every six months for the USFS. The first female chief of the USFS, Gail Kimbell, was appointed in 2007.

Lawsuits would be filed against the Department of Labor in 1974, the Department of Energy in 1976, and the U.S. Geological Survey in 1978. The plaintiffs won in each instance with some combination of back pay, promotions, and compliance programs as the result. NASA even began admitting women as astronauts with the 1978 class, apparently now believing that they too could possess "the right stuff." Physicist Sally Ride, a member of that 1978 class, would become the first American woman in space in 1983 [1, 10].

In 1978, chemist Mollie Glesier filed suit against the Lawrence Berkley Laboratory after having been laid off and not re-hired although men less qualified then her were getting jobs. Although the American Chemical Society supported her financially, she lost her case in 1980. Another national laboratory, Lawrence Livermore National Laboratory was also sued although much later. In 1998, chemist Mary Singleton filed a class-action lawsuit that made allegations including unequal pay and lack of promotional opportunities. This suit took five years but was successful and resulted in significant changes in the leadership of the Laboratory [1].

In the 1980s, another issue that had remained beneath the surface for years surfaced – that of sexual harassment. Jenny Jew, an anatomy professor at the University of Iowa, was denied a promotion to full professor because of allegations of a sexual relationship with the department chair. Outraged at this allegation, she sued the university in 1985. In spite of a full-fledged battle with the university's lawyers, she won the case in 1990, got her promotion, damages, back pay and benefits and recovered her attorneys' fees. Five Yale University students filed a case in federal court in 1988 for damages related to retaliation when they refused faculty members' demands for sexual favors. Although their case was unsuccessful, notice had been served to universities that this type of behavior was unacceptable on the part of the faculty [1].

The resulting fight for promotional opportunities and equal pay hadn't yet been resolved, however. In 1986, mathematician Jenny Harrison was denied tenure. Hired as the third woman in the mathematics department at the University of California, Berkeley, Harrison filed suit in 1989 and asked to see the evidence leading to her tenure denial as part of the court case. Because the U.S. Supreme Court had decided in a different case that universities had to provide confidential personnel files in a discrimination case, the administration at Berkeley became involved and Harrison got not only tenure but a full professorship. Other institutions took notice as well [1].

6.5   Increasing Number of Women Pursuing Science Education in the 1970s

In the physical sciences, women earning PhDs increased to almost 11% of the total in 1979 from just over 5% in 1969, from 12.2% of the total earning masters in 1969 to 18.6% in 1979, and from 14.3% in 1969 at the bachelor's level to 22.6% in 1979. Similar progress, or better, was experienced in other scientific fields including environmental sciences, mathematics and computer sciences, life sciences, and social sciences as shown in Table 6.1 [11]. An intriguing outcome of the efforts to recruit women students into the sciences was that those increasing numbers of women expected to take classes from, perform research under, and be advised by women faculty. Old sexist attitudes and behaviors towards women faculty were no longer going to be acceptable [1].

6.6   The 1980s

In the early 1980s, Americans began worrying about whether or not the country was on an equal footing with technologically-advanced Japan. As a result of this international competitiveness, more focus was placed on engineers and technology in the U.S., in the hopes of keeping America economically robust. The pipeline for engineers and scientists began to be discussed, and the number of women in science and technology careers began to receive significant focus and recognition [12].

The National Science Foundation (NSF) Authorization and Science and Technology Equal Opportunities Act of 1980 was passed to include women and minorities as problem solvers to deal with the now recognized issues of environ-

Table 6.1 Scientific degrees (%) for women 1969 to 1979

Academic field	PhD		Master's		Bachelor's	
	1969	1979	1969	1979	1969	1979
Physical sciences	5.4	10.9	12.2	18.6	14.3	22.6
Environmental sciences	4.0	9.0	7.9	17.4	10.0	22.8
Mathematics and computer sciences	5.1	14.9	23.3	26.7	36.6	35.9
Life sciences (biological/agricultural sciences)	14.1	22.9	22.6	32.3	24.4	37.1
Social sciences	11.5	25.6	22.6	33.5	36.5	42.9

ment, food shortages, and areas affected by affirmative action. Both Jewel Plummer Cobb and the OOS had invested significant time and resources to ensure that the Act passed [1, 12, 13, 14]. The act said:

> ... it is the policy of the United States to encourage men and women, equally, of all ethnic, racial, and economic backgrounds to acquire skills in science, engineering and mathematics, to have equal opportunity in education, training, and employment in scientific and engineering fields, and thereby to promote scientific and engineering literacy and the full use of the human resources of the Nation in science and engineering. [15]

The Visiting Professorships for Women (VPW) program was created by the passage of this act and the associated funding for NSF. This program underwrote visiting professorships for 25–30 women for six to 24 months at an institution of their choosing. The host institution got the overhead from the grant which made it enticing for them. The visiting woman professor was expected to spend 30 percent of her time encouraging women students and scientists at the host institution. Not only did the VPW program provide the means for women to serve as role models at a broad array of institutions over the course of the 17 years of the program, but it also provided fodder for the NSF to introduce other programs for women scientists [1].

Yet in spite of this rhetoric, the Reagan administration cut science education funding in the early 1980s. The falloff in the rate of increase in the number of women pursuing scientific careers is evidenced by the flattening of the women graduates by the 1990s. This outcome is attributed, in part, to the cuts in federal funding and the Reagan's administration significantly reduced emphasis on affirmative action [11].

Warning signs abounded of the importance of education. The 1983 report *A Nation at Risk*, produced by the U.S. National Commission on Excellence in Education, stated that the U.S.'s "once unchallenged preeminence in commerce, industry, science, and technological innovation is being overtaken by competitors throughout the world." It further stated that "We have even squandered the gains in student achievement made in the wake of the Sputnik challenge." Many recommendations were made as to ways in which the K-12 experience should be improved to prepare students for the information age [1, 16].

The so-called Neal Report which is actually NSB 86-100, a report of the National Science Board titled *Undergraduate Science, Mathematics and Engineering Education*, recommended expansion in funding for science education [1, 17].

A very unexpected source of support for women in science began in 1987. Clare Boothe Luce's will bequeathed money to her husband's Henry Luce Foundation to create the Clare Boothe Luce Program. The purpose of this new program was "to encourage women to enter, study, graduate, and teach" in science, mathematics and engineering. Although she had not shown interest in women in science during her life, after her death, the $70 million in funds was used to endow assistant professorships in the physical sciences and engineering at private, many of them Catholic, universities. Fourteen colleges and universities benefit in perpetuity from her largesse. Others can apply for grants from the Program. The Program grew to include biological sciences and many women who received funding through the program were the first female faculty at their institutions. By 2020, the Program can say it "has become one of the single most significant sources of private support for women in science, mathematics and engineering in Higher Education in the United States." It has supported more than 2500 women [1, 18].

Congress became convinced by 1987 that based on manpower projections for scientists and engineers that showed significant shortfalls by 2006, something needed to be done. A law was passed creating a *Task Force on Women, Minorities and the Handicapped in Science and Engineering* to examine the current status of those groups in the targeted fields and to coordinate existing federal programs to promote their education and employment in science and engineering. The Task Force Report was issued in 1989 and concluded that non-traditional engineers and scientists faced barriers in both promotion and progression in their careers [12, 19].

The latest national imperative to get women and minorities into technology was reflected in a 1988 report [12]:

> *If compelled to single out one determinant of US competitiveness in the era of the global, technology-based economy, we would have to choose education, for in the end people are the ultimate asset in global competition.... But an especially important further step will be to extend the pool from which the pipeline draws by bringing into it more women, more racial minorities, and more of those who have not participated because of economic, social, and educational disadvantage... Thus not only is providing a better grounding in math and science for all citizens a matter of making good on the American promise of equal opportunity. It is a pragmatic necessity if we are to maintain our economic competitiveness.*

Additional legislative and regulatory actions in the 1980s and into the 1990s helped improve the workforce for women in general. Specifically, the Equal Employment Commission issued regulations in 1980 that defined sexual harassment as a form of sex discrimination, thus prohibited under the Civil Rights Act of 1964. U.S. Supreme Court rulings in the 1980s and 1990s further clarified the situations constituting sexual harassment [4]. Women scientists, who tended to be more isolated in work environments because of their fewer numbers, now had greater recourse for some of the more blatant behavior that might be experienced.

6.7   The 1990s

In the 1990s, the U.S. was now focused on ways to remain globally competitive with the entire world, not just Japan [12]. More initiatives, task forces, studies, and conferences occurred to further examine what often boils down to the phrase "Why so Few?" Why aren't more women pursuing scientific careers? Many issues had been identified and, in some scientific fields, women had made significant progress. Why that progress had not been uniform and how the percentage and number of women could be further increased in all areas of science were issues that continued to be examined [11].

6.8   Powre

The NSF replaced the VPW with the Professional Opportunities for Women in Research and Education (POWRE) in 1997. Through this program, junior women scientists trying to attain tenure were provided with $75,000 grants. The selection process was very selective and 600 grants were made over the four years of the program's existence [1].

6.9   The Commission on the Advancement of Women and Minorities in Science, Engineering and Technology

The Commission on the Advancement of Women and Minorities in Science, Engineering and Technology (CAWMSET), established by Congress in 1998, again examined the issues and potential remedies associated with the low participation of women, minorities, and persons with disabilities in science, engineering, and technology careers. The Commission's Report *Land of Plenty* (September 2000) identified issues and made recommendations with regard to precollege education, access to higher education, professional life, public image, and nationwide accountability [20].

6.10   A Study on the Status of Women Faculty in Science at MIT

The March 1999 the Massachusetts Institute of Technology (MIT) Faculty Newsletter contained a bombshell felt at many university campuses around the U.S. The women science faculty at MIT felt marginalized and excluded from a significant role in their departments. The marginalization increased as women progressed in their careers and resulted in differences in salary, space, awards, resources, and responses to outside offers between men and women faculty with women receiving less in all categories despite professional accomplishments equal to those of their male colleagues. The findings repeated through successive generations of women faculty and the percentage of women faculty in the School of Science had not changed significantly for 10, or perhaps, 20 years [21, 22].

6.11   Science in the Twenty-First Century

The U.S. continued to need all the technical talent it could find. Women were more prevalent in the twenty-first century in the sciences than they ever have been. However, scientific careers still attracted fewer women than many other professional fields. Many issues needed to be resolved in order for the number of young women pursuing scientific careers to increase significantly. Another national call to arms had been issued by the CAWMSET [20]:

> As we enter the twenty-first century, U.S. jobs are growing most rapidly in areas that require knowledge and skills stemming from a strong grasp of science, engineering, and technology... business leaders are warning of a critical shortage in skilled American workers that is threatening their ability to compete in the global marketplace. Yet, if women, underrepresented minorities, and persons with disabilities were represented in the U.S. science, engineering, and technology (SET) workforce in parity with their percentages in the total workforce population, this shortage could largely be ameliorated... Now, more than ever, the nation needs to cultivate the scientific and technical talents of all its citizens, not just those from groups that have traditionally worked in the SET fields. It is apparent that just when the U.S. economy requires more SET workers, the largest pool of potential workers continues to be isolated from SET careers... if the nation is willing to make the investment ...[such an investment] yields approximately four or five to one returns in economic benefits to the nation. If, on the other hand, the United States continues failing to prepare citizens from all population groups for participation in the new, technology-driven economy, our nation will risk losing its economic and intellectual preeminence.

6.12   Advance

In 2000, the National Science Foundation decided to initiate what it called the ADVANCE program: Organizational Change for Gender Equity in STEM Academic Professions. The goal of the program is institutional transformation through systemic change and it encourages adaptation and partnership. Universities receive large grants, $3-5 million, and use that money to set up programs helping women attain equity at all levels. Top officials at universities must agree to set up these types of programs. Seventy-two proposals were received in response to the first call for proposals and eight institutions received the first awards in the fall of 2001. From 2001 through 2016, 160 institutions in 47 states, the District of Columbia and Puerto Rico have received $270 million through the ADVANCE program [1, 23, 24].

6.13   Lawrence Summers: 2005

And yet, issues that many women thought had been resolved kept arising. The President of Harvard University, Lawrence Summers, in January of 2005 raised three issues that have haunted women in science throughout the twentieth and twentyfirst centuries. Summers stated that women wouldn't work the long hours needed to succeed in the science field because of childcare responsibilities. Then, he stated that genetic differences, i.e., innate abilities, between boys and girls drive different interests in scientific careers not socialization. And thirdly, he stated that there wasn't sex bias in appointments in academic institutions. His comments set off a tremendous furor that led to his resignation and the appointment of the first female president of Harvard University since its founding in 1636 [1, 25–27].

6.14   Nelson Diversity Surveys

In 2000, chemist Dr. Donna Nelson of the University of Oklahoma starting gathering data on women and minorities at U.S. research universities after her students asked her about collecting data on minorities in chemistry. She has since published the Nelson Diversity Surveys documenting the lack of women and minorities on science faculty around the country. As she says, "Progress for female and minority faculty at research universities, produced from past attempted solutions combined, has been too slow. If significant progress is to be made within the next couple of decades, new and totally different approaches to solving problems facing women and minority faculty will be needed."

The data from her 2004 report which reflects B.S. candidates from 2000 and faculty data for 2002 except for chemistry and astronomy which are for 2003 were as follows for science fields excluding engineering:

Dr. Nancy Hopkins, who led the MIT School of Science study, is quoted in the report: "Who can look at these numbers and not say that we as a faculty have failed – failed our students, failed our institutions, and most of all, failed our nation?" Dr. Marye Anne Fox, the Chancellor of North Carolina State University, said: "It was discouraging to know that went I went to (the University of) Texas in 1976, I was the second woman in a faculty of about 50, and when I left in 1998, they were again hiring a second woman" [1, 28].

6.15   National Research Council: Committee on Science, Engineering and Public Policy

As a result of Lawrence Summers's comments, the National Research Council established a committee to prepare a report on women on science faculties. Their report, published in 2007 was titled *Beyond Bias and Barriers: Fulfilling the Potential of Women in Academic Science and Engineering*. In summary, it said:

> The United States economy relies on the productivity, entrepreneurship, and creativity of its people. To maintain its scientific and engineering leadership amid increasing economic and educational globalization, the United States must aggressively pursue the innovative capacity of all its people – women and men. However, women face barriers to success in every field of science and engineering; obstacles that deprive the country of an important source of talent. Without a transformation of academic institutions to tackle such barriers, the future vitality of the U.S. research base and economy are in jeopardy.
>
> Beyond Bias and Barriers *explains that eliminating gender bias in academia requires immediate overarching reform, including decisive action by university administrators, professional societies, federal funding agencies and foundations, government agencies, and Congress. If implemented and coordinated across public, private, and government sectors, the recommended actions will help to improve workplace environments for all employees while strengthening the foundations of America's competitiveness.* [1, 29]

6.16 President's Council of Advisors on Science and Technology

On February 7, 2012, the President's Council of Advisors on Science and Technology released a report titled *Engage to Excel: Producing One Million Additional College Graduates with Degrees in Science, Technology, Engineering, and Mathematics.* Because of the perceived need for one million more graduates in science, technology, engineering and mathematics (STEM) over the next decade in order for the U.S. to maintain its global preeminence in science and technology, the report recommended changes in undergraduate education, especially in the first two years of college, in order to retain students pursuing a STEM education. Such improved retention would be the most inexpensive way to fill the expected gap. The report said that a:

> large and growing body of research indicates that STEM education can be substantially improved through a diversification of teaching methods. These data show that evidence-based teaching methods are more effective in reaching all students – especially the 'underrepresented majority' – the women and members of minority groups who now constitute approximately 70% of college students while being underrepresented among students who receive undergraduate STEM degrees (approximately 45%). This underrepresented majority is a large potential source of STEM professionals.

Yet, again, another call to arms for increasing the number of women and minorities in the STEM professions was issued [30].

6.17 Yale Gender Bias Study

Seeking to address the continuing disparity between the number of women receiving their PhDs in science and the number of women hired as junior faculty, Yale faculty and students undertook an analysis of gender bias in hiring. Their technique was to take the same resume and randomly assign it a male or female name. They then informed the hiring recipient that the applicant had completed his/her B.S. and was looking to be hired as a lab manager. The results demonstrated consistent gender biases against hiring the female applicant – whether the individual doing the hiring analysis was male or female. The categories evaluated were competence, hireability, likability and the extent to which the individual would be willing to mentor the applicant. Although the female applicant scored higher on likability, the male applicant was more likely to be hired and mentored and more likely to be deemed competent. In addition, the potential salary that would be offered to the male candidate was significantly higher than the female candidate [31].

6.17.1 Why So Few Still?

More than half of today's student body at colleges and universities are female – the percentage reported in the fall of 2019 was 56%. Yet, women are still not anywhere near these percentage levels as students for certain areas of the sciences or on the faculty in certain science departments. Surprisingly, for those of us who have spent the past forty years encouraging women to pursue STEM careers, many of the same issues that kept women from pursuing undergraduate and graduate degrees in the sciences and being hired as faculty in the sciences departments have not changed in forty or fifty years. The data in Table 6.2 from the Nelson Diversity Study and Tables 6.3, 6.4, and 6.5 tell the story [32, 33].

Table 6.2 Nelson diversity study – gender distribution of BS recipients versus role models

	% Female		% Male	
	Students	Faculty	Students	Faculty
Chemistry	47.3	12.1	52.7	87.9
Math	48.2	8.3	51.8	91.7
Computer science	27.7	10.6	72.3	89.4
Astronomy	32.7	12.4	67.3	87.6
Physics	21.4	6.6	78.6	93.4
Economics	32.3	11.5	67.7	88.5
Political science	50.1	23.5	49.9	76.5
Sociology	70.2	35.8	29.8	64.2
Psychology	76.5	33.5	23.5	66.5
Biological sciences	58.4	20.1	41.6	79.9

Table 6.3 Scientific bachelor's degrees (%) for women 1979, 2000, 2012, 2017

Academic field	1979	2000	2010	2017
Physical sciences	22.6	41.1	41.3	45.9
Environmental sciences	22.8	–	–	–
Earth, atmospheric and ocean sciences	–	40.0	39.3	38.9
Mathematics and computer sciences	35.9	32.7	25.6	25.1
Biological and agricultural sciences	37.1	55.8	57.8	60.3
Social and behavioral sciences	42.9	63.0	62.5	64.6

Table 6.4 Scientific master's degrees (%) for Women 1979, 2000, 2012, 2015

Academic field	1979	2000	2010	2017
Physical sciences	18.6	34.63	37.5	36.3
Environmental sciences	17.4	–	–	–
Earth, atmospheric and ocean sciences	–	38.1	47.0	43.3
Mathematics and computer sciences	26.7	35.6	30.6	33.2
Biological and agricultural sciences	32.3	52.9	56.2	58.1
Social and behavioral sciences	33.5	61.3	64.7	65.4

Table 6.5   Scientific PhD (%) for women 1979, 2000, 2012, 2015

Academic field	1979	2000	2010	2017
Physical sciences	10.9	25.0	31.7	30.9
Environmental sciences	9.0	–	–	–
Earth, atmospheric and ocean sciences	–	28.7	40.5	42.7
Mathematics and computer sciences	14.96	21.8	25.7	24.9
Biological and agricultural sciences	22.9	42.9	52.2	51.8
Social and behavioral sciences	25.6	55.5	57.6	60.1

Many steps have been taken although the playing field is not level yet for women or minorities. More work is needed in the twenty-first century in order to utilize all the talent available to its highest potential and address the scientific challenges that humanity faces. May the U.S. truly learn to embrace the skills and capabilities of all members of its population and, maybe, in the not too distant future, women will reach parity with men in their pursuit and success in scientific careers.

***

KEY WOMEN OF THIS PERIOD

***

6.18   Stephanie Kwolek: Chemist (1923–2014)

Stephanie Kwolek (Fig. 6.1), the fourth woman inducted into the National Inventor's Hall of Fame (1995), is best known for her invention of Kevlar™, the lightweight yet very strong polymer used in bulletproof vests and many other products. Kwolek spent 40 years with DuPont during which time she obtained 16 patents for a variety of groundbreaking materials and devised new processes in polymer chemistry.

Kwolek showed an early interest in science – and in fashion design. Maybe it is not surprising that much of her scientific research thus centered on fibers. Intending to pursue a career as a doctor, Kwolek graduated with a BS in chemistry from what is now Carnegie-Mellon University in 1946. She accepted a position as a chemist in the rayon department with DuPont planning to save the money she needed to attend medical school. But she became so interested in the research in which she was involved that she decided medical school was no longer in her future.

One of her early projects was the preparation of polymers that formed Lycra™ spandex fibers, the stretchy material used in athletic wear. Another notable product to emerge from her work is Nomex, which is fire-resistant and now commonly used in protective gear worn by firefighers. In the course of looking for a strong, lightweight fiber that could be used in radial tires, Kwolek discovered what became Kevlar™. Kevlar™ is used in bulletproof vests, helmets for the military, ropes, fiber-optic cables, aircraft parts, brake linings and canoes. It is five times stronger than the same weight of steel.

Fig. 6.1 Stephanie
Kwolek. (Courtesy of
Science History Institute)

Among her many awards, Kwolek received the National Medical of Technology
in 1996, only the second individual woman to do so. She was the fourth woman
inducted into the National Inventors Hall of Fame (1995). Kwolek has been inducted
into the Hall of Fame of Delaware Women and the National Women's Hall of
Fame [34].

6.19   Jewel Plummer Cobb: Biologist (1924–2017)

Jewel Plummer Cobb became interested in science in ninth grade when her biology
teacher put a microscope in front of her. Her parents had always encouraged her
interest in education and had introduced her to the wonders of science as well as
strong women, particularly African-American women.

Cobb entered the University of Michigan in 1941 but stayed for just three semesters because of the racist treatment. There was no support system for black students,
the dormitories were segregated, black students were not allowed in the Pretzel Ball
or Beer Parlor, and women couldn't walk in the front door of the men's union building. So she transferred to Talladega College, founded by the American Missionary
Society just after the abolition of slavery, and graduated with a BS in biology in

1944.[1] An MS and PhD, both in cell physiology and both from New York University, followed in 1947 and 1950.

Cobb's research over the years has primarily been related to cancer causes and treatment including extensive study of melanin – the brown or black pigment that colors skin and its ability to shield human skin from ultraviolet rays. Her melanin studies involved examination of melanoma, a form of skin cancer. As her research evolved, so did her career, as she moved up the academic ladder to become President of California State University at Fullerton in 1981.

In her academic administrative positions, Cobb initiated a number of programs to encourage ethnic minorities and women to pursue careers in the sciences. She was the first minority appointed to the National Science Board (1974) and one of the youngest. Through her position on the NSB, she devoted much time and effort to ensuring the passage of the National Science Foundation Authorization and Science and Technology Equal Opportunities Act of 1980 [1, 34–36].

6.20 Nancy Roman: Astronomer (1925–2018)

If you enjoy puzzles, science or engineering may be the field for you, because scientific research and engineering is a continuous series of solving puzzles. It is also a continuous process of learning new things, whether you discover them or study the work of others.

"The Mother of Hubble" Nancy Roman (Fig. 6.2) organized an astronomy club for her friends when she was eleven years old. They met once a week to learn about constellations. That love of astronomy lasted throughout her life. That love was fostered by both of her parents. Her geophysicist father answered her scientific questions. Her mother took her for walks during the day showing her plants and birds. The walks at night were to observe constellations and aurora.

Roman earned her BA in astronomy at Swarthmore College in 1943 and her PhD from the University of Chicago in 1949. She worked at the Yerkes Observatory and the U.S. Naval Research Observatory after obtaining her PhD. While attending a scientific conference a few months after NASA was formed in 1959, she was approached and asked if she knew anyone who would like to set up a program in space astronomy at NASA. She interpreted that as an offer to apply – which she did. Roman became the first Chief of Astronomy in the Office of Space Science at NASA Headquarters and the first woman to hold an executive position at NASA.

Roman's responsibilities at NASA included plans to observe objects in space by using rockets and satellite observatories. She was involved in the Orbiting Solar Observatories and her programs led to the successful *Viking* probes to collect data from the Mars planetary surface. The last program in which she was involved before her retirement was the Hubble telescope. Her contributions to the early planning of the Hubble Space Telescope, the establishment of its program structure, and the lob-

---

[1] Talladega College, in Talladega, Alabama, is a private historically black college founded in 1867.

Fig. 6.2 Nancy Roman.
(Courtesy NASA)

bying effort for Congress to fund it led to her being called the "Mother of Hubble." Her many awards include being named a Fellow of the American Astronautical Society, honorary degrees, having an asteroid named for her and being memorialized as a minifig (a LEGO® minifgure) in the "Women of NASA" LEGO® series [34, 37–39].

6.21   Vera Rubin: Astronomer (1928–2016)

Vera Rubin's research was focused on the study of galaxies: their movement, internal rotation, and distribution. Because of her work, scientists now believe that up to 90 percent of the universe may be composed of dark matter – invisible to the naked eye.

Rubin was attracted to the stars as a child and built a small telescope through which she endeavored to photograph the moon. She said,

> By about age 12, I would prefer to stay up and watch the stars than go to sleep. I started learning. I started going to the library and reading. But it was initially just watching the stars from my bedroom that I really did. There was just nothing as interesting in my life as watching the stars every night.

Rubin attended Vassar College, in part, because Maria Mitchell had taught astronomy there. After receiving her BA in 1948, she earned an MA at Cornell in 1951, turning down an offer to study at Harvard to be with her husband at Cornell. Her master's thesis examined the evidence for bulk rotation in the universe by studying 108 galaxies.

The family moved to Washington, D.C. and Rubin studied at Georgetown University receiving her PhD in 1954 with a dissertation that showed galaxies clumped together instead of being randomly distributed. After ten years as a research astronomer and assistant professor of astronomy at Georgetown (and raising four children), Rubin took up observational astronomy. Her time at Mount Palomar was groundbreaking – America's preeminent observatory was reserved for male use only as there were no toilet facilities for women. She was the first woman officially permitted to observe there.

Rubin's colleague at the Department of Terrestrial Magnetism (DTM) of the Carnegie Institution, Kent Ford, and Rubin began to study the systematic motion of galaxies again in the mid-1970s. What they found was that a large group of galaxies including our own Milky Way are moving rapidly with respect to the rest of the universe. Called the Rubin-Ford effect, this theory appears to support the clumpiness theory of matter distribution in the universe.

Rubin received many awards for her work including election in 1981 to the National Academy of Sciences. She actively promoted women in astronomy and wrote a children's book on astronomy, to encourage young girls to study science. All four of the Rubin children became scientists as well [34, 40, 41].

6.22 Jean Sammet: Computer Scientist (1928–2017)

Named a fellow in 2001 of the Computer History Museum "For her contributions to the field of programming languages and its history," Jean Sammet wanted to attend the Bronx High School of Science, but couldn't because girls weren't allowed. She did pursue her mathematics interest, however, earning her bachelor's degree in mathematics from Mount Holyoke College and her master's in mathematics from the University of Illinois. From 1955 to 1958, she worked at Sperry Gyroscope supervising the scientific programming group, the company's first, and teaching graduates programming courses part-time at Adelphi College. During the years she managed software development for the Army Signal Corps at Sylvania Electric Products, she also was a key member of the committee that developed COBOL, the business application language used around the world.

In 1961, she went to work for IBM, where she directed the development of the language used for symbolic mathematics, FORMAC. Later, she led IBM's work on the Ada programming language (named for Ada Bryon Lovelace). From 1974–1976, she served as the first female president of the Association of Computing Machinery (ACM).

Sammet was an authority on the history of programming languages.[2] As she said:

From childhood on I hated to throw papers away. As I became an adult this characteristic merged with my interest in computing history. As a result I created important files and docu-

---

[2] Jean Sammet played a pivotal role in the nomination of Admiral Grace Murray Hopper for the National Medal of Technology. The author submitted that nomination. Jean Sammet told me that she would help me on one condition: that I did everything that she told me to do and that I wrote everything that she told me to write. I complied with her request.

ments of my own, and became concerned with having other people publish material on their important work so the facts (rather that the myths) would be known publicly.

Sammet was elected to the National Academy of Engineering in 1977 and received an honorary degree from Mount Holyoke College. Her other awards include the Ada Lovelace Award from the Association for Women in Computing and the Computer Pioneer Award from the Institute of Electrical and Electronics Engineers (IEEE) Computer Society [42, 43].

6.23   Tu Youyou: Pharmaceutical Chemist (1930–)

The recipient of a shared Nobel Prize in Physiology or Medicine in 2015, Tu Youyou (Fig. 6.3) was the first mainland Chinese scientist to receive a Nobel Prize.[3] Youyou developed a compound to treat malaria, based on traditional Chinese medical texts, that has saved millions of lives. Her accomplishment is even more outstanding since she did so without a doctorate, a medical degree or training abroad.

Growing up with four brothers in a family that valued education, Youyou decided to study medicine after contracting tuberculosis at age 16 and taking a two-year break to recover. She wanted to find cures for diseases like the one she had experienced. She studied at Beijing Medical College and became a pharmacologist classifying medicinal plants, extracting their ingredients and identifying their chemical structures. When she graduated in 1955, she went to work at the Academy of Traditional Chinese Medicine and stayed for the rest of her career.

In 1967, Youyou became part of Project 523 – the search to find a treatment for chloroquine-resistant malaria. In 1969, she was appointed head of the project. After researching malaria in situ for three years on Hainan Island, Youyou and her team returned to Beijing. By the time they returned, more than 240,000 compounds had been tested and none had been effective.

Youyou and her team reviewed the medical texts of the Zhou, Quing and Han dynasties. Around 400 A.D., the texts referenced sweet wormwood which had been used to treat intermittent fevers, which the team knew was a symptom of malaria. In 1971, her team isolated a compound that appeared to battle the parasite that caused malaria. She discovered that boiling the wormwood compromised the compound's capabilities. Thus, she used an ether-based solvent on the wormwood instead. During tests on monkeys and mice, the compound was 100% effective.

The first human guinea pigs were Tu Youyou and two of her team members. All 21 patients in Hainan Province who received the test drug recovered. The active ingredient was identified as artemisinin. After her results were published in English in 1979, she was invited to present her findings to the World Health Organization, the World Bank, and the United Nations. After many years, the World Health Organization recommended artemisinin as the first line of defense against malaria.

---

[3] The others Nobel Prize recipients were William C. Campbell and Satoshi Omura.

Fig. 6.3 Tu Youyou.
(Courtesy of Bengt
Nyman)

In 2011, the Lasker Medical Foundation in awarding Youyou its Clinical Medical Research Award called artemisinin "arguably the most important pharmaceutical intervention in the last half-century."

On receiving the Nobel Prize, Youyou said "Every scientist dreams of doing something that can help the world" [44].

6.24 Fran Allen: Computer Scientist (1932–)

The first woman IBM fellow, Fran Allen (Fig. 6.4) specializes in compilers, compiler optimization, programming languages, and parallelism. Her early computer work culminated in algorithms and technologies that are the basis for the theory of program optimization and are widely used throughout the industry. In the early 1980s, she founded the Parallel Translation Group (PTRAN) to study compiling for parallel machines. This group was recognized as one of the top research groups in the world dealing with parallelization issues.

Allen started as a teacher having earned a degree from Albany State Teachers College (now the State University of New York at Albany). She taught algebra, geometry, trigonometry, and practical math to farm children at a small rural high school in Peru, New York where she had gone to school. To be a fully certified teacher in New York requires a master's degree, so Allen went to the University of Michigan for hers, intending to return to teaching. However, she needed to pay off her debt and accepted a job at IBM. Instead of being temporary, however, her employment at IBM lasted for more than 40 years.

Allen is a member of the National Academy of Engineering (elected in 1987), a fellow of the Institute of Electrical and Electronics Engineers (IEEE), a fellow of the Association for Computing Machinery (ACM), and an elected fellow of the American Academy of Arts and Sciences (AAAS). Allen received the Pace University School of Computer Science and Information Systems award for Leadership and Service in Technology. This award recognizes individuals within the technology field that serve people and are committed to community service and education. In 2006, Allen became the first woman to receive ACM's Turing Award in its forty-year history. The citation reads: *For pioneering contributions to the theory and practice of optimizing compiler techniques that laid the foundation for modern optimizing compilers and automatic parallel execution.* She has received a number of honorary degrees as well [45–51].

6.25   Elinor Ostrom: Economist (1933–2012)

The first woman to win the Nobel Prize in Economics, Elinor Ostrom (Fig. 6.5) credited the debate team and the swimming team with laying the foundation for her career.[4] She learned not only how to work in teams but also that there were at least two sides to every policy issue and how to establish and critique the arguments that

---

[4] She shared the Nobel Prize with Oliver E. Williamson.

Fig. 6.5 Elinor Ostrom.
(Courtesy Holger Motzkau
Wikipedia)

each side developed. She worked during her years in college at the University of California, Los Angeles and earned all three of her degrees there in political science. She received her PhD in 1965.

Ostrom's work related to how communities manage common resources. A political scientist, her approach to the issue of community management of "the commons" – resources such as pastures, woods, lakes and other bodies of water, and fisheries – equitably and sustainably over the long term, differed from approaches taken by economists. Instead of looking at the issue hypothetically, she looked at reality. She conducted field studies and she carefully analyzed what her research had uncovered. She demonstrated that community common resources could be managed without central authority or privatization. She developed eight principles that can be used by communities to successfully manage "the commons." There is also an Ostrom law, named for her, that states "A resource arrangement that works in practice can work in theory."

Ostrom and her husband Vincent Ostrom spent their careers at Indiana University where she became a Distinguished Professor. Among her many honors in addition to the Nobel Prize, Ostrom was elected to the National Academy of Sciences, received an honorary degree, and was named one of *Time* magazine's "100 Most Influential People in the World" for 2012. She was active until her death – with her last article published on the day that she died in 2012 [52, 53].

6.26   Sylvia Earle: Marine Biologist (1935–)

Called "Her Deepness", a "Living Legend" and a "Hero for the Planet", oceanographer, author, lecturer, and marine biologist Sylvia Earle was encouraged by her parents in her love of nature. However, when it was time for her to go to college, they supported her major in biology but wanted her to get her teaching credentials and learn how to type "just in case." Earle received her B.S. from Florida State University in 1955 and her master's degree in botany from Duke University in 1956. Her laboratory became the Gulf of Mexico as her thesis was a detailed study of algae in the Gulf. She received her PhD from Duke University in 1966.

After serving as the resident director of the Cape Haze Marine Laboratories in Sarasota, Florida, she moved to Massachusetts and served as both a research scholar at the Radcliffe Institute and research fellow at the Farlow Herbarium, Harvard University. In 1976, she moved to California where she became a research biologist and curator at the California Academy of Sciences and a fellow of botany at the Natural History Museum, University of California, Berkeley.

The love of the sea was also calling during these years. In 1970, Earle lived in an underwater chamber for fourteen days with four other oceanographers as part of the government-funded Tektite II Project. This was the first all-female team and she was its leader. All told, she has led more than 50 expeditions and logged more than 7000 h under water.

With the advent of SCUBA equipment, the study of marine biology was dramatically changed. Earle was one of the first researchers to use the SCUBA equipment and observe plant and animal habitats and life beneath the sea and was able to identify many new species of each. In addition, she set the unbelievable record of free diving to a depth of 1250 ft.

Earle and her former husband recognized the limitations of SCUBA and set up a company in 1981 to build submersible craft that could dive deeper than humans in SCUBA gear. The design for the *Deep Rover* submersible was sketched on a napkin. It continues to operate as a mid-water machine capable to going to ocean depths of up to 3000 ft.

Earle was the first woman to serve as the chief scientist for the National Oceanic and Atmospheric Administration (NOAA). She has published books and hundreds of scientific papers and other publications on marine life. She has been at the helm of ocean exploration for more than four decades. Earle is a dedicated advocate of public education regarding the importance of oceans as an essential environmental habitat. To promote ocean education, in 2015, The LEGO Group issued a Dr. Sylvia Earle LEGO® set including a minifig of Dr. Earle as well as a deep sea exploration vessel, deep sea submarine, deep sea helicopter, deep sea starter set and a deep sea SCUBA scooter.

Earle has received more than 100 national and international honors including the Netherlands Order of the Golden Ark, Australia's International Banksia Award, and medals from the Royal Geographic Society, the National Wildlife Federation, and the Philadelphia Academy of Sciences. A tireless advocate for the Earth's oceans, she has been inducted into the National Women's Hall of Fame [34, 54, 55, 56].

Fig. 6.6 Johnnetta
B. Cole. (Courtesy of
NASA)

6.27   Johnnetta B. Cole: Anthropologist (1936–)

The first black woman president in Spelman College history, Johnnetta B. Cole
(Fig. 6.6) began her college education at age 15 at Fisk University.[5] She transferred
to Oberlin College and earned her undergraduate degree in sociology in 1957. She
earned her master's and PhD degrees in anthropology from Northwestern University,
doing her dissertation field work in Liberia, West Africa. She then pursued an academic career. While at Washington State University, she founded one of the first
black studies programs in the U.S. At the University of Massachusetts at Amherst,
she was instrumental in establishing the W.E.B. Du Bois Department of African-
American Studies. During her time at Hunter College, Cole was a professor of
anthropology and also served as director of the Latin American and Caribbean
Studies Program. She assumed the presidency of Spelman College in 1987 and
would stay a decade.

When she left Spelman, Cole taught at Emory University as the Presidential
Distinguished Professor of Anthropology, Women's Studies and African-American
Studies. She then assumed the presidency of Bennett College in Greensboro, North
Carolina from 2002–2007. Bennett is the only other historically black college and

---

[5] Spelman College is a private black liberal arts college for women in Atlanta, Georgia. Fisk
University is an historically black university in Nashville, Tennessee.

university (HBCU) dedicated to educating black women other than Spelman. After briefly retiring, she served as the Director of the Smithsonian's National Museum of African Art from 2009–2017.

Cole has received numerous awards and more than 40 honorary degrees. She served on the boards of directors of Home Depot, Merck and Coca-Cola. She was the first woman elected to Coke's board. She was the first African American to chair the board of United Way of America [57, 58].

6.28   Mary Dell Chilton: Plant Biotechnologist (1939–)

In 1983, Mary-Dell Chilton (Fig. 6.7) led the research team that produced the first transgenic plants.[6] As such, she is considered one of the founders of modern plant biotechnology and the field of genetic engineering in agriculture. After groundbreaking efforts at the University of Washington and Washington University, she established one of the world's leading industrial biotechnology agricultural programs at Ciba-Geigy AG (today Syngenta AG). Her team has worked to produce crops with higher yields, and resistance to pests, disease and adverse environmental conditions (such as drought).

Chilton was elected to the National Academy of Sciences in 1985. The recipient of numerous other awards including the 1985 Rank Prize in Nutrition and the 2013 World Food Prize, she was inducted into the National Inventors Hall of Fame in 2015. Her name is now on the building in the Research Triangle Park in North Carolina where Distinguished Science Fellow Chilton worked.

Dr. Chilton's B.S. and Ph.D. degrees are in chemistry from the University of Illinois Urbana-Champaign. She said "My career in biotechnology has been an exciting journey and I am amazed to see the progress we have made over the years. My hope is through discoveries like mine and the discoveries to follow, we will be able to provide a brighter and better future for the generations that follow us" [59, 60, 61].

6.29   Ada E. Yonath: Crystallographer (1939–)

The 2009 Nobel Laureate in Chemistry, Ada E. Yonath (Fig. 6.8) says she was curious as a child.[7] Her family emigrated from Poland to Israel where she grew up in impoverished conditions. Her parents encouraged her in her pursuit of an education. Yonath did her bachelor's and master's work in chemistry, biochemistry and bio-

---

[6] Transgenic means genetically modified – DNA from an unrelated organism has been artificially introduced into the original organism.

[7] The Nobel Prize was shared with Venkatraman Ramakrishnan and Thomas A. Steitz.

Fig. 6.7 Mary-Dell
Chilton

Fig. 6.8 Ada Yonath –
Weizmann Institute of
Science

physics at Hebrew University of Jerusalem. Her PhD is from the Weizmann Institute and she did postdoctoral work at what is today Carnegie-Mellon as well as at MIT.

In 1970, Yonath returned to the Weizmann Institute where she stayed for the remainder of her career. There she established the first biological crystallography lab in Israel. Her project was to figure out the process of protein biosynthesis for which she needed to determine the three-dimensional structure of the ribosome. The ribosome is found within all living cells and it links amino acids together. Her work, which evolved into a center for macromolecular assemblies, led to a greater understanding of the workings of some of the most widely prescribed antibiotics. Through dogged determination in the face of a large number of skeptics, Yonath was able to prove the feasibility of ribosome crystallography. Through this process, Yonath developed new techniques used around the world in structural biology laboratories. In 2000 and 2001, the first complete three-dimensional structures of the ribosome were published.

Yonath's Nobel Prize citation reads "for studies of the structure and function of the ribosome." She professes excitement that the ribosome has now come into the public eye and that her granddaughter invited her to come to her kindergarten class to talk about ribosomes! Since she has curly hair, a saying in Israel now is "Curly hair means a head full of ribosomes" [62].

6.30   Christiane Nüsslein-Volhard: Developmental Biologist (1942–)

German Nobel Laureate Christiane Nüsslein-Volhard (Fig. 6.9) knew she wanted to pursue a career in biology by the time she was 12. After a few changes of major and a change of university, she completed her degree in biochemistry at the University of Tübingen. After a PhD and post-doctoral studies in which she became interested in genetics, choosing to work with the *Drosophila* fly, she became established at the Max-Planck Institute for the rest of her career.

Her work involved implementing techniques to uncover development of the fly embryos at the earliest stages. These methods and techniques revolutionized developmental genetics. She and her group discovered much of what is known about how fertilization takes place and how organisms begin their lives. In addition to her work with the *Drosophila* fly, she has worked with zebrafish – the embryos are transparent which facilitates experimentation.

The citation for the 1995 Nobel Prize in Physiology or Medicine reads "for their discoveries concerning the genetic control of early embryonic development."[8] Her numerous other recognitions include the Albert Lasker Award, election to the U.S. National Academy of Sciences, many honorary doctorates, and having an asteroid named for her [36, 63, 64].

---

[8] The Nobel Prize was shared with Edward B. Lewis and Eric F. Wieschaus.

Fig. 6.9 Christiane
Nüsslein-Volhard.
(Courtesy of Rama)

Fig. 6.10 JoAnn Cram
Joselyn. (Courtesy
Colorado Women's Hall of
Fame)

6.31   JoAnn Cram Joselyn: Astrogeophysicist (1943–)

The first woman to earn a doctorate from the astrogeophysics program at the
University of Colorado – Boulder, JoAnn Cram Joselyn (Fig. 6.10) spent her career
with NOAA. Her focus was on predicting space weather including solar flares and
sunspots as well as their impacts on terrestrial and space communications.

Joselyn was the first woman to serve as the Secretary General of the International Association of Geomagnetism and Aeronomy. She was the first woman and the first American to serve as the Secretary General of the International Union of Geodesy and Geophysics. Among her many awards, Joselyn has been inducted into the Colorado Women's Hall of Fame [65, 66].

6.32   Jocelyn Bell Burnell: Astrophysicist and Astronomer (1943–)

Dame Jocelyn Bell Burnell discovered the pulsar (Fig. 6.11) in 1967 while a graduate student at the University of Cambridge, England. Although she did not receive the Nobel Prize for her discovery (her advisor did), she has received many other recognitions over her lifetime. In 2018, one such recognition was the $3 million Special Breakthrough Prize in Fundamental Physics which cited her "detection of radio signals from rapidly spinning, super-dense neutron stars" as well as her "lifetime of inspiring scientific leadership." Burnell used all of the prize money to establish the Bell Burnell Graduate Scholarship Fund to provide financial assistance to female, minority and refugee populations in order to enable them to pursue physics research.

Fig. 6.11 Chart from which Jocelyn Bell Burnell identified the first Pulsar. (Courtesy Wikipedia (Billthom))

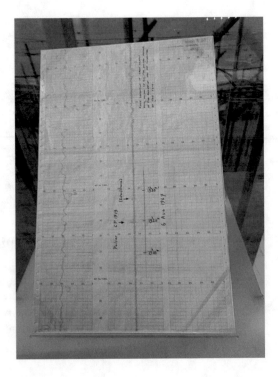

Burnell was encouraged in her love of astronomy by her parents and by the staff at the Armagh Observatory near her home in Belfast, Northern Ireland. After earning a degree in physics from Glasgow University, she pursued her graduate studies at the University of Cambridge. Once she had assisted in the construction of a radio telescope, her job was to analyze the data that it generated. She recognized the pattern of the first pulsar which is shorthand for pulsating radio star. Her discovery opened up new areas of astronomy. She published her results and received her PhD in 1968.

She married soon after and accompanied her husband to various locations around Great Britain, working part-time as she raised their son. She also pursued her commitment to social justice and to her religious faith. After her divorce, she returned to astronomy full time, becoming the third woman physics professor in the United Kingdom when she accepted a position at the Open University. Burnell has received a number of honors including selection as a Fellow of the Royal Society as well as the Royal Society of Edinburgh. She received the Herschel Medal of the Royal Astronomical Society and gave the Lise Meitner Lecture at the Technical University of Vienna [67–70].

6.33   Shirley Jackson: Physicist (1946 –)

The first African-American woman to receive a PhD from MIT, theoretical physicist Shirley Jackson (Fig. 6.12) is now the President of Rensselaer Polytechnic Institute (RPI). Jackson's area of expertise is particle physics – the branch of physics that predicts the existence of subatomic particles and the forces that bind them together.

Jackson was encouraged in her interest in science by her father who helped her with projects for her science classes. She took accelerated math and science classes in high school and graduated as valedictorian. At MIT, she was one of less than twenty African-American students on campus, the only African American studying physics, and one of about 43 women in the freshmen class of 900. After obtaining her BS at MIT, she opted to stay for her doctoral work in order to encourage more African-American students to attend the institution. She completed her dissertation and obtained her PhD in 1973.

After postdoctoral work at prestigious laboratories in the U.S. and abroad, Jackson joined the Theoretical Physics Research Department at AT&T Bell Laboratories in 1976. She served on the faculty at Rutgers University from 1991 to 1995 and then became the first woman and African-American Chairman of the U.S. Nuclear Regulatory Commission. In 1999, she became the first African-American and first woman President of RPI.

Her numerous honors and awards include induction into the National Women's Hall of Fame and the Women in Technology International Hall of Fame, the Thomas Alva Edison Science Award, and the CIBA-GEIGY Exceptional Black Scientist. Jackson actively promotes women in science [34, 35, 71, 72].

Fig. 6.12 Shirley Jackson.
(Courtesy of Wikipedia)

6.34   Candace Pert: Neuroscientist and Pharmacologist (1946–2013)

Like Jocelyn Bell Burnell, Candace Pert's discovery of opiate receptors while she was a graduate student, resulted in accolades for her advisor, but not for her. Pert dropped out of college to get married and raise a family. While working as a cocktail waitress, she spoke with a customer who was an assistant dean at Bryn Mawr. With the help of that assistant dean, Pert applied to Bryn Mawr. Pert graduated with a BA in biology in 1970 and became a graduate student in the pharmacology department at Johns Hopkins University.

It was there that Pert began explorations of the body's regulatory mechanisms for neurotransmitters. Her next project was the search for opiate receptors; at that time opiate receptors were believed to exist but that existence had not been proven. These receptors are where endorphins bind in the brain. Pert used radioactive drugs to identify the receptor molecules and published her first results in the prestigious magazine *Science* in 1973. Her research also identified that opiate receptors were present during fetal development. Pert received her PhD in 1974.

In 1978, Pert's advisor received the Lasker Award for these discoveries; Pert did not. She continued working at Johns Hopkins until 1982 when she became chief of the section on brain chemistry at the National Institute of Mental Health. Here, she led the team that discovered peptide T, a compound of amino acids that blocks binding and infection of the human immunodeficiency virus (HIV). She continued her work on peptides at the company she founded and later joined the faculty at Georgetown University.

Pert was also a significant advocate of the mind-body connection and its emergence as a legitimate area of medical research. She was affectionately called "The Mother of Psychoneuroimmunology", and "The Goddess of Neuroscience." Pert published more than 250 articles [34, 73, 74].

6.35 Shirley Malcom: Zoologist/Ecologist (1946–)

A leader in the effort to improve the quality of and access to education in the STEM fields, Dr. Shirley Malcom (Fig. 6.13) obtained her bachelor's degree in zoology from the University of Washington, her master's also in zoology from UCLA and her PhD in ecology from Pennsylvania State University. She taught at the university level before becoming a research associate at the American Academy for Arts and Sciences.

Surprised by the paucity of women and minorities in the student body and on the faculty in the STEM fields around the country, she began work in 1977 with the National Science Foundation to address that issue. She has become a worldwide champion of STEM education for women and using science and technology (S&T) to solve problems in daily lives. Malcom has served or serves on many policy-making bodies. She was a member of the National Science Board as well as President Clinton's Committee of Advisors on Science and Technology. She served as co-chair of the Gender Advisory Board of the UN Commission on S&T for Development and Gender InSITE. The recipient of numerous honorary degrees and other awards, Malcom received the Public Welfare Medal of the National Academy of Sciences, its highest honor, in 2003 [75, 76].

Fig. 6.13 Shirley Malcom. (Courtesy NASA)

6.36   Linda Buck: Biologist (1947–)

The Nobel Laureate in 2004 in Physiology or Medicine "for their discoveries of odorant receptors and the organization of the olfactory system," Linda Buck (Fig. 6.14) grew up with parents who liked to solve puzzles and invent.[9] She believes that sowed the seeds for her later interest in discovery as did their encouragement that she could do anything she wanted. Buck's interest in science was sparked by an immunology class she took in college such that she decided to pursue biology. Her undergraduate work at the University of Washington was followed by graduate work in the Microbiology Department at the University of Texas Medical Center.

While doing postdoctoral work at Columbia University, Buck read a 1985 paper on how odor detection might work. She became fascinated by the concept as to how the olfaction system could distinguish between 10,000 smells. She began her work on the olfaction system in 1988. In 1991, she co-published the first paper on the identification of odorant receptors. She continued her olfactory system work at Harvard and later at the University of Washington.

The recipient of many awards in addition to the Nobel Prize, Buck says, "As a woman in science, I sincerely hope that my receiving a Nobel Prize will send a message to young women everywhere that the doors are open to them and that they should follow their dreams" [77].

6.37   Temple Grandin: Animal Scientist (1947–)

A professor of animal science at Colorado State University, Temple Grandin (Fig. 6.15) is an expert on animal behavior. She is also a spokesperson for autism. Half of the cattle in the U.S. are handled in facilities that she designed. Animals in Canada, Mexico, Europe, Australia and New Zealand are also housed in Grandin-designed facilities. She is a proponent of humane treatment of livestock for slaughter.

On the advice of her high school science teacher, Grandin developed her 'hug box' or 'squeeze machine' to provide herself with stress relief. It was later adapted for use with animals. Grandin did not speak until she was three and a half years old but was able to be mainstreamed by the time she was in kindergarten. Her science teacher was an important mentor and sparked her interest in pursing a scientific career. She has authored more than 60 peer-reviewed scientific papers. Grandin earned her bachelor's degree in psychology from Franklin Pierce College (Today Franklin Pierce University), and her master's and PhD degrees, both in animal science, from Arizona State University and the University of Illinois, Urbana-Champaign, respectively.

Her first book, *Emergence: Labeled Autistic,* was the first inside narrative written about autism. She was also one of the first adults to disclose that she was autistic, thus publicizing the condition. In 2010, *Time* magazine named her one of the 100

---

[9] The Nobel Prize was shared with Richard Axel.

Fig. 6.14 Linda Buck.
(Courtesy Wikipedia)

Photo by Rosalie Winard

Fig. 6.15 Temple Grandin. (Courtesy Colorado Women's Hall of Fame)

most influential people in the "Heroes" category. Her life story was told in an HBO movie. Her many awards include honorary degrees and induction into the National Women's Hall of Fame [78, 79].

6.38   Françoise Barré-Sinoussi: Virologist (1947–)

Awarded the Nobel Prize in Physiology or Medicine in 2004 for the "discovery of human immunodeficiency virus" Françoise Barré-Sinoussi grew up observing plants and animals on her vacations.[10] She began as a volunteer at the Institut Pasteur in Paris in 1975. She not only became a full-time employee there; it is there that she would spend the rest of her career.

Torn between medicine and sciences, Barré-Sinoussi opted for a degree in natural sciences as she thought the educational process for such would be less expensive and thus less of a burden on her family than a medical degree. As she was completing her baccalaureate in natural sciences from the University of Paris, Barré-Sinoussi became interested in the possibility of a research career and was eventually able to get a position at the Institut Pasteur. That position evolved quickly into her PhD research and she received her PhD from the Faculty of Sciences at the University of Paris in 1974. After a post-doctoral research fellowship at the NIH, Barré-Sinoussi returned to the Institut Pasteur.

In 1982, she began working to discover the cause of the new epidemic that would later be named acquired immunodeficiency syndrome (AIDS). She isolated the virus and characterized it with the first results reported in *Science* magazine in May 1983. HIV would continue to be her focus for the remainder of her career involving much work and travel to resource-limited countries in Africa and Asia.

Barré-Sinoussi has received numerous other awards in addition to the Nobel Prize including the National Order of the Legion of Honor (the highest award in France), honorary degrees, and the Prize of the French Academy of Sciences [80, 81].

6.39   Flossie Wong-Staal: Molecular Biologist (1947–)

Virologist and molecular biologist, Flossie Wong-Staal was the first scientist to clone HIV, a major step in proving that HIV was the cause of AIDS and thus leading to effective treatments of the disease.

Her family fled mainland China and relocated to Hong Kong where she completed her primary and secondary schooling. Wong-Staal's teachers in Hong Kong encouraged her to pursue a scientific education even though no woman in her family had ever studied science or even worked outside of the home. Wong-Staal attended the University of California, Los Angeles, earning her undergraduate and graduate degrees in bacteriology and molecular biology. Her postdoctoral work was conducted at the University of California, San Diego (UCSD). Wong-Staal then joined the National Cancer Institute (NCI). While at NCI, she and her team identified HIV as the cause of AIDS and completed the genetic mapping of the virus. Wong-Staal

---

[10]The Nobel Prize was shared with Harald zur Hausen and Luc Montagnier.

said about working with HIV, "Working with this virus is like putting your hand in a treasure chest. Every time you put your hand in, you pull out a gem." She later returned to UCSD, and continued her work on retroviruses.

Elected to the Institute of Medicine of the U.S. National Academies, Wong-Staal retired from UCSD and served as Chief Scientific Officer for iTherX, formerly Immusol, a biopharmaceutical company she co-founded while at UCSD. The company is working on improved drugs for the treatment of hepatitis C. Among her many honors, Wong-Staal has been inducted into the National Women's Hall of Fame [82, 83].

6.40 Lydia Villa-Komaroff: Molecular and Cellular Biologist (1947–)

Lydia Villa-Komaroff (Fig. 6.16) overcame active discouragement and discrimination to become a successful molecular and cellular biologist. Interested in science from age nine, she attended the University of Washington. There she changed her major from chemistry to biology after being told that women didn't belong in chemistry. Moving to the Washington, DC area to be with the man who would become her husband, she applied to Johns Hopkins University to finish her bachelor's degree but was told that they were not accepting female students. Undeterred, she completed her BS in biology at Goucher College. Graduate school followed at MIT,

Fig. 6.16 Lydia Villa-Komaroff. (Courtesy Wikipedia)

where she completed her PhD in cell biology in 1975. During her time at MIT, Villa-Komaroff co-founded the Society for Advancement of Chicanos and Native Americans in Science to address the paucity of these groups in the sciences.

Villa-Komaroff pursued an academic career. She was a professor at the University of Massachusetts Medical School, Harvard University, and Northwestern University before moving into private industry. Through her work, molecular biology has been transformed from a field of study into an essential tool. The focus of her research work has been on using methods of recombinant DNA to address a variety of fundamental questions in different areas.

Villa-Komaroff is a Fellow of AAAS and AWIS. She is also an advocate of diversity in STEM careers. She says, "there is nothing quite like finding something truly novel and potentially important. If you are doing cutting-edge science, then that is a rare feeling, since many times things don't work… seeing the scientists I trained go on to successful careers of their own is very rewarding" [84–86].

6.41   Elizabeth Blackburn: Biochemist (1948–)

The first Australian woman Nobel Laureate, Elizabeth Blackburn (Fig. 6.17) shared the 2009 Nobel Prize in Physiology or Medicine "for the discovery of how chromosomes are protected by telomeres and the enzyme telomerase."[11] She was fond of animals and nature from an early age. Growing up in Tasmania, she read the biography of Marie Curie, written by Marie's daughter Eve, and knew by the time she was a teenager that she wanted to do science. After moving to Australia her final year of high school, she enrolled at the University of Melbourne. She completed her bachelor's and her master's there in biochemistry. Her PhD is from the University of Cambridge in England.

Love and fortune would intervene and her postdoctoral work was at Yale University. Later, Blackburn would serve on the faculty at first the University of California at Berkeley and then the University of California at San Francisco. Her work showed that telomeres, the caps at the end of each chromosome in a cell nuclei, has a particular DNA. Further, she found that the DNA prevents the chromosomes from breaking down. In addition, she and Carol Greider discovered the enzyme telomerase which produces the DNA found in the telomere. She later served as President of the Salk Institute for Biological Studies.

Blackburn has received many awards and recognition for her work in addition to the Nobel Prize. These include the National Academy of Sciences Award in Molecular Biology, Foreign Associate of the National Academy of Sciences, Fellow of the AAAS, Fellow of the Royal Society, and the Albert Lasker Award for Medical Research. She is also a Fellow of the Australian Academy of Science [87, 88].

---

[11] The Nobel Prize was shared with Carol Greider and Jack W. Szostak.

6.42   Anita Borg: Computer Scientist (1949–2003)

Anita Borg earned her BS, MS, and PhD (1981) degrees in computer science from
Courant Institute of Mathematical Sciences, New York University. Early in her
career, Borg was the lead designer and co-implementer of a fault tolerant micropro-
cessor, message-based UNIX system. The system provided users the ability to run
programs that would automatically recover from hardware failures. She designed
and built the first software system for generating and analyzing extremely long
address traces. The knowledge gained from this effort was used in the development
of Digital Equipment Corporation's Alpha technology. Later, she designed and
managed the implementation of Mecca, a web-based email system used by thousands of people.

Borg was known for much more in the computer industry than her significant
technical accomplishments. In 1987, she founded the "Systers" email list linking
technical women in computing when email was in its infancy. Borg founded the
Grace Hopper Celebration of Women in Computing in 1994. After joining Xerox
in1997, she created a center to find ways to apply information technology to assure
a positive future for the world's women. The Institute for Women and Technology
(renamed the Anita Borg Institute after her death) researches, develops, and deploys
useful, usable technology in support of women's communities.

Borg received many honors and recognitions including induction into the Women
in Technology International Hall of Fame, the Melitta Benz Women of Innovation

and Invention Award, the Pioneer Award from the Electronic Frontier Foundation, the August Ada Lovelace Award from the Association of Women in Computing, and the Heinz Award for Technology, the Economy and Environment. She was a fellow of ACM. She holds two patents. In 1999, she was the presidential appointee (by President Clinton) to the Commission on the Advancement of Women and Minorities in Science, Engineering and Technology [89–98].

6.43   Maria Klawe: Computer Scientist (1951–)

The first woman to serve as the president of Harvey Mudd College, computer scientist Maria Klawe began in that role in 2006. Canadian by birth, she grew up in Canada and Scotland. She attended the University of Alberta, earning her BSc as well as a PhD in mathematics there. Klawe started a second PhD in computer science at the University of Toronto but became a faculty member before completing the degree. She spent time in industry, at the University of British Columbia and at Princeton University before being named president of Harvey Mudd. Her research interests include functional analysis, discrete mathematics and human-computer interaction.

Klawe has actively recruited female students and female faculty to Harvey Mudd and speaks across the country about the programs and approaches she uses to ensure a more diverse institution. When she arrived, the students and faculty at Harvey Mudd, which offers nine engineering, science and mathematics-based majors, was about 30% female. By 2017, 45% of the students and 40% of the faculty were female.

Klawe has received many awards including the Lifetime Achievement Award from the Canadian Association of Computer Science and was ranked 17 on Fortune's 2014 list of the World's Greatest Leaders. She is a Fellow of the Canadian Information Processing Society, the AAAS, the American Mathematical Society, and the Association for Women in Mathematics [99–101].

6.44   Ingrid Daubechies: Physicist and Mathematician (1954–)

Belgian mathematician Ingrid Daubechies was a child prodigy. Instead of counting numbers to get to sleep when she was very young – she did multiplication tables. By age six, her parents said she was already familiar with mathematical concepts including cone and tetrahedron. She completed her undergraduate work in physics at Vrije Universiteit Brussel in 1975 and her PhD in theoretical physics in 1980 from Free University in Brussels.

She is best known for the orthogonal Daubechies wavelets and the biorthogonal CDF (Cohen-Daubechies-Feauveau) wavelets. Her work has been a refinement of the Fourier technique used to decrease the size of digital photos and movies so they take up less kilobytes without a significant reduction in quality of detail. Her work has fundamentally changed image and signal processing. She has also developed image processing techniques that have helped validate the authenticity of Renaissance art.

As a professor at Duke University, Daubechies lists her research interests as wavelet theory, signal processing, machine learning, computational geometry, and time-frequency analysis. Her numerous awards include a MacArthur Fellowship, National Academy of Engineering, National Academy of Sciences, and the first woman to receive the National Academy of Sciences Award in Mathematics. She has been elected a foreign member of the Royal Netherlands Academy of Arts and Sciences and has been named a Baroness by the King of Belgium [102, 103].

6.45   Susan Solomon: Chemist (1956–)

Susan Solomon (Fig. 6.18), senior scientist with the NOAA, has received significant recognition for her work in explaining the cause of the Antarctic ozone hole including the National Medal of Science (1999) and the Weizmann Women & Science Award (2002). The National Medal of Science citation reads: "For key scientific insights in explaining the cause of the Antarctic Ozone 'hole' and for advancing the understanding of the global ozone layer; for changing the direction of ozone research through her findings; and for exemplary service to worldwide policy decisions and to the American public." Solomon carried out key work theorizing that chemical reactions involving manmade chlorine could be responsible for the observed ozone depletion in the Antarctic. She served as the leader of the National Ozone Expeditions to the Antarctic in 1986 and 1987. An Antarctic glacier was named in her honor in recognition of her work. Her follow-up work theorizes that major volcanic eruptions can also lead to the loss of atmospheric ozone.

Solomon grew up in Chicago and became interested in science by watching the undersea adventures of Jacques Cousteau. In high school, she took third place in a nationwide science fair. She attended the Illinois Institute of Technology where she received her BS in chemistry with high honors. She received her MS and PhD (1981) degrees in chemistry at the University of California at Berkeley. After many years as a research scientist at NOAA, Solomon joined the faculty at MIT. From 2012–2016, she was the Ellen Swallow Richards Professor of Atmospheric Chemistry & Climate Science at MIT and later assumed a different endowed professorship. Her book, *The Coldest March*, published in 2001, explores the ill-fated expedition of Captain Robert Scott to the South Pole. The book theorizes that exceptionally cold temperatures thwarted the expedition, killing the entire team, and not incompetence on the part of Captain Scott.

Solomon has received numerous other awards including the James B. MacElwane Medal of the American Geophysical Union, the Ozone Award from the United Nations Environment Programme, and the 1999 Carl-Gustaf Rossby Award from the American Meteorological Society. She is a member of the National Academy of

Sciences, a foreign associate of the French Academy of Sciences, and a foreign member of the Academia Europaea. She has been inducted into the National Women's Hall of Fame [104–110].

6.46   Fran Arnold: Chemical Engineer, Bioengineer and Biochemist (1956)

On the way to her 2018 Nobel Prize in Chemistry, Fran Arnold (Fig. 6.19) didn't follow a conventional education or career path.[12] She moved out of her parents' house in Pittsburgh, Pennsylvania at 17 but did eventually end up as an undergraduate at Princeton University. At a time when no one had heard of a semester abroad, she spent a year in Italy. She graduated with a bachelor's degree in mechanical and aerospace engineering and went to work in the solar energy field. When funding for renewable energy dried up during the Reagan administration, she went back to school earning a master's in chemical engineering and a PhD in biochemistry, both from the University of California at Berkeley. In 1986, she became a faculty member at Cal Tech.

The initial intent of her research was to use emerging DNA technology to design new enzymes. As she discovered how difficult this was, she decided to change course and to let nature optimize using its own technique: evolution. In 1993, she demonstrated how using directed selection led to the development of new enzymes. She reflected that evolution is "a force of nature that has led to the finest chemistry

---

[12] The Nobel Prize was shared with George P. Smith and Sir Gregory P. Winter.

Fig. 6.19 Fran Arnold.
(Courtesy Bengt Nyman)

of all time." This new technique of using directed evolution to design new enzymes has applications that range from pharmaceuticals to renewable energy.

Arnold's many honors in addition to the Nobel Prize include the National Medal of Technology and Innovation and induction into the National Inventors Hall of Fame. She was the first woman to receive the Draper Prize from the National Academy of Engineering. She was the first woman to be elected to all three of the National Academies in the U.S. – the National Academy of Engineering, the National Academy of Sciences and the National Academy of Medicine. The first woman to win the Millennium Technology Prize, Arnold was the first American woman to win the Nobel Prize in Chemistry. She said "If you're going to change the world, you've got to be fearless" [111, 112].

6.47 Donna Strickland: Physicist (1959–)

When it was announced that Canadian Donna Strickland (Fig. 6.20) would be the third woman to win the Nobel Prize in Physics in 2018 (after Marie Curie and Maria Goeppert-Mayer), the world was surprised that she didn't have a Wikipedia bio and that she wasn't a full professor at the University of Waterloo. She said she had never gotten around to filling out the paperwork for the full professorship; shortly thereafter she did. The Wikipedia oversight was corrected almost immediately.

The work for which she won the Nobel Prize was for the "method of generating high-intensity, ultra-short optical pulses."[13] Also referred to "chirped pulse amplifi-

---

[13] The Nobel Prize was shared with Arthur Ashkin and Gérard Mourou.

cation," the work that Strickland did while a graduate student at the University of Rochester has many applications including for corrective eye surgery.

Strickland grew up in Ontario, Canada and became interested in laser and electro-optics. She received her Bachelor of Science in engineering physics from McMaster University and her masters and PhD from the University of Rochester. After employment at the National Research Council of Canada, at Lawrence Livermore National Laboratory and at Princeton University, she joined the faculty at the University of Waterloo in 1997. Strickland was the first full-time female professor in physics there. Since receiving the Nobel Prize, she was appointed a Companion of the Order of Canada, one of Canada's highest civilian honors [113, 114].

6.48   Carol Greider: Molecular Biologist (1961–)

The co-recipient of the 2009 Nobel Prize in Physiology or Medicine, Carol Greider (Fig. 6.21) overcame dyslexia as a child through creativity and persistence.[14] The Nobel Prize was for her role in discovering telomerase, an enzyme that is critical to the aging process and to the growth of cancer cells; the knowledge of which is critical for the purposes of medical research.

Growing up, Greider was put into remedial classes because of her dyslexia. She thought of herself as stupid but ultimately developed a technique for reading that involved memorization. This would serve her well in her history and biology classes.

---

[14]The Nobel Prize was shared with Elizabeth Blackburn and Jack W. Szostak.

Fig. 6.21 Carol Greider.
(Courtesy Johns Hopkins
School of Medicine)

After studying marine ecology at the College of Creative Studies at the University of California Santa Barbara, Greider worked in a lab at the Max Planck Institute before deciding to go to graduate school. Only two of the universities she applied accepted her – Cal Tech and the University of California, Berkeley. When she met Elizabeth Blackburn at UC, Berkeley, her decision was made. She would study telomeres with Blackburn. At age 23, before she earned her PhD, she made the discovery for which she would win a Nobel Prize. Today, she is at Johns Hopkins working with telomeres and telomerase.

Greider is concerned about women's underrepresentation still in science. She encourages women to combine work and family saying, "find a way to do it. There's more than one way," just as she found out for herself when she was learning to read [115, 116].

6.49   May-Britt Moser: Psychologist and Neuroscientist (1963–)

Norwegian Nobel Laureate May-Britt Moser shared the 2014 Nobel Prize in Physiology or Medicine for "discoveries of cells that constitute a positioning system in the brain."[15] Her research, jointly with her then husband Edvard Moser, uncovered cells in the hippocampus of the brain that are important for positioning. They also discovered that these cells communicate with other cells to effectively form a navigation system.

---

[15] The Nobel Prize was shared with John O'Keefe and Edvard I. Moser.

Moser grew up on a farm in western Norway. Encouraged by her family and her teachers, she enrolled at the University of Oslo, in close proximity to where two of her sisters lived. She and her husband-to-be decided to both study psychology so they could learn about the brain. For graduate work, they both studied neuroscience at the same University and worked with rats in a water maze. And, they both got their PhDs in neuroscience from the University of Oslo as well. Moser was determined to do this and she said that around this time she realized how insistent she could be – although she was nice and polite, when she really wanted something, no one could stop her. And, she managed to get her PhD with two small children.

Postdoctoral work at the University of Edinburgh followed as well as research time at University College London. Their research time was cut short, however, when a position opened at the Norwegian University of Science and Technology (NTNU) in Trondheim. The Mosers were able to convince the University to hire them both and to provide them with the lab in which to do their research. She became a full professor of neuroscience in 2000.

Moser continues her work at NTNU where she is the Founding Director of the Centre for Neural Computation and Co-Director of the Kavli Institute for Systems Neuroscience. She is also the Co-Founder of the Centre for the Biology of Memory [117–119].

References

1. M. Rossiter, *Women Scientists in America: Forging a New World since 1972*, vol 3 (The Johns Hopkins University Press, Baltimore, 2012)
2. M.W. Rossiter, *Women Scientists in America: Before Affirmative Action 1940–1972* (The Johns Hopkins University Press, Baltimore, 1995)
3. Association for Women in Computing – Home. https://www.awc-hq.org/home.html. Accessed 17 Apr 2020
4. S. Tobias, *Faces of Feminism: An Activist's Reflections on the Women's Movement* (Westview Press, Boulder, 1997)
5. Milestones: 1972. https://www.eeoc.gov/eeoc/history/35th/milestones/1972.html. Accessed 12 Apr 2020
6. Milestones: 1970, EEOC. https://www.eeoc.gov/eeoc/history/35th/milestones/1970.html. Accessed 17 Apr 2020
7. Milestones: 1973, EEOC. https://www.eeoc.gov/eeoc/history/35th/milestones/1973.html. Accessed 17 Apr 2020
8. B. J. Love (ed.), *Feminists Who Changed America 1963–1975* (University of Illinois Press, Urbana and Chicago, 2006)
9. R.B. Marimont, NIH Mathematician, https://www.washingtonpost.com/archive/local/2004/03/31/obituaries/799f6b0d-6a78-4b5d-92b2-ea52fd7eaf31/. Accessed 17 Apr 2020
10. First woman named head of U.S. Forest Service. http://www.nbcnews.com/id/16598898/ns/us_news-environment/t/first-woman-named-head-us-forest-service/#.XpoiVMhKjIU. Accessed 17 Apr 2020
11. E. Babco, *Professional Women & Minorities: A Total Human Resources Data Compendium*, Commission on Professionals in Science & Technology, 13th edn. Apr 2000, ISSN: 0190-1796

12. J.C. Lucena, "'Women in Engineering'" a history and politics of a struggle in the making of a statistical category, in *Proceedings of the 1999 International Symposium on Technology and Society – Women and technology: Historical, Societal, and Professional Perspectives*, pp. 185–194, New Brunswick, 29–31 July 1999

13. S. 568 (96th): National Science Foundation Authorization and Science and Technology Equal Opportunities Act. https://www.govtrack.us/congress/bills/96/s568. Accessed 17 Apr 2020

14. National Science Foundation, *Women, Minorities, and Person With Disabilities in Science and Engineering: 1996*, Arlington, 1996, (NSF 96–311)

15. Science and Engineering Equal Opportunity Act, Section 32(b), Part B of P.L. 96–516, 94 Stat. 3010, as amended by P.L. 99–159

16. A Nation at Risk. https://www2.ed.gov/pubs/NatAtRisk/risk.html. Accessed 17 Apr 2020

17. National Science Board, NSB 86-100. https://www.nsf.gov/nsb/publications/1986/nsb0386.pdf. Accessed 17 Apr 2020

18. Clare Boothe Luce Program – Program Aims. https://www.hluce.org/programs/clare-boothe-luce-program/. Accessed 17 Apr 2020

19. NSPE Women in Engineering Task Force, *The Glass Ceiling & Women in Engineering* (NSPE Publication, Alexandria 1992)

20. Congressional Commission on the Advancement of Women and Minorities in Science, Engineering and Technology Development, *Land of Plenty*, Sept 2000

21. *The MIT Faculty Newsletter*, vol. XI, no. 4, Mar 1999, Special Edition, "A Study on the Status of Women Faculty in Science at MIT," http://web.mit.edu/fnl/women/women.html. Accessed 25 Dec 2001

22. R. Wilson, An MIT Professor's suspicion of Bias leads to a new movement for academic women, *The Chronicle of Higher Education*, 3 Dec 1999

23. ADVANCE: Organizational Change for Gender Equity in STEM Academic Professions. https://www.nsf.gov/funding/pgm_summ.jsp?pims_id=5383. Accessed 18 Apr 2020

24. ADVANCE Brochure, https://www.nsf.gov/ehr/Materials/ADVANCEBrochure.pdf. Accessed 18 Apr 2020

25. Evans, Clay, "Science needs women," *Daily Camera*, Boulder, 13 Feb 2005

26. L.H. Summers, Remarks at NBER Conference on Diversifying the Science & Engineering Workforce, 14 Jan 2005. www.president.harvard.edu/speeches/2005/nber.html

27. Why women are poor at science, by Harvard president. https://www.theguardian.com/science/2005/jan/18/educationsgendergap.genderissues. Accessed 18 Apr 2020

28. D.J. Nelson, Jan 2004 (revised 2005 and 2007), A National Analysis of Diversity in Science and Engineering Faculties at Research Universities. http://drdonnajnelson.oucreate.com/diversity/briefings/Diversity%20Report%20Final.pdf. Accessed 18 Apr 2020

29. https://www.nap.edu/catalog/11741/beyond-bias-and-barriers-fulfilling-the-potential-of-women-in. Accessed 18 Apr 2020

30. President's Council of Advisors on Science and Technology Release Report Outlining Undergraduate Education Initiative, American Institute of Physics, 14 Feb 2012. https://www.aip.org/fyi/2012/presidents-council-advisors-science-and-technology-release-report-outlining-undergraduate. Accessed 23 Apr 2020

31. Science faculty's subtle gender biases favor male students, C.A. Moss-Racusin, J.F. Dovidio, V.L. Brescoll, M.J. Graham, J. Handelsman, Proceedings of the National Academy of Sciences of the United States of America, 9 Oct 2012. https://www.pnas.org/content/109/41/16474. Accessed 23 Apr 2020

32. E. Pollack, *The Only Woman in the Room: Why Science Is Still a Boys' Club* (Beacon Press, Boston, 2015)

33. The State of U.S. Science and Engineering 2020. https://ncses.nsf.gov/pubs/nsb20201. Accessed 18 Apr 2020

34. P. Proffitt (ed.), *Notable Women Scientists* (The Gale Group, Detroit, 1999)

35. S.A. Ambrose, K.L. Dunkle, B.B. Lazarus, I. Nair, D.A. Harkus, *Journeys of Women in Science and Engineering: No Universal Constants* (Temple University Press, Philadelphia, 1997)

36. B. F. Shearer, B. S. Shearer (eds.), *Notable Women in the Life Sciences* (Greenwood Press, Westport, 1996)
37. N. Roman (1925–2018) – Astronomer/"Mother of Hubble. https://solarsystem.nasa.gov/people/225/nancy-roman-1925-2018/. Accessed 12 May 2020
38. N.G. Roman, Happy 90th Birthday Nancy! https://women.nasa.gov/nancy-grace-roman-2/. Accessed 13 May 2020
39. N. Roman. https://en.wikipedia.org/wiki/Nancy_Roman. Accessed 13 May 2020
40. B. F. Shearer, B. S. Shearer (eds.), *Notable Women in the Physical Sciences* (Greenwood Press, Westport, 1997)
41. V. Rubin, The Astronomer Who Brought Dark Matter to Light Tim Childers, 11 June 2019. https://www.space.com/vera-rubin.html
42. J. Sammet, 2001 Fellow, Computer History Museum. https://computerhistory.org/profile/jean-sammet/. Accessed 19 Apr 2020
43. J.E. Sammet. https://en.wikipedia.org/wiki/Jean_E._Sammet. Accessed 19 Apr 2020
44. T. Youyou. https://www.nobelprize.org/womenwhochangedscience/stories/tu-youyou. Accessed 19 Apr 2020
45. F. Allen, IBM Fellow, to give UCSD Regents' Lecture 16 Jan 1997. www.sdsc.edu/SDSCwire/v2.25/fran_allen_news.html. Accessed 24 Apr 2002
46. Kimberley A. McGrath, *Who's Who in Technology*, 7th edn (Gale Research, Inc., New York, 1995)
47. D. Gürer, Grace hopper conference banquet speech on pioneering women in computing, 19 Sept 1997. www.acm.org/women/speech.html. Accessed 24 Apr 2002
48. K. Colborn, C. Willard, Fran Allen of IBM: pushing the limits of computing, *Diversity/Careers in Engineering & Information Technology*, 23 Apr 2001
49. F. Allen (1932–2000) Fellow Award Recipient. www.computerhistory.org/exhibits/hall_of_fellows/allen/. Accessed 24 Apr 2002
50. F.E. Allen. https://en.wikipedia.org/wiki/Frances_E._Allen. Accessed 19 Apr 2020
51. Frances ("Fran") Elizabeth Allen. https://amturing.acm.org/award_winners/allen_1012327.cfm. Accessed 19 Apr 2020
52. E. Ostrom: Facts https://www.nobelprize.org/prizes/economic-sciences/2009/ostrom/facts/. Accessed 13 May 2020
53. E. Ostrom. https://en.wikipedia.org/wiki/Elinor_Ostrom. Accessed 13 May 2020
54. National Geographic, Explorers: Bio – Sylvia Earle. http://www.nationalgeographic.com/explorers/bios/sylvia-earle/
55. TED: Ideas worth spreading, Sylvia Earle – Oceanographer. https://www.ted.com/speakers/sylvia_earle
56. C. Chin, Review of Mission blue documentary about Dr. Sylvia Earle, Aug 26 2014. http://protecttheoceans.org/wordpress/?p=1459
57. About Johnnetta B. Cole, Ph.D. https://www.spelman.edu/about-us/office-of-the-president/past-presidents/johnnetta-cole. Accessed 19 Apr 2020
58. J.B. Cole. https://en.wikipedia.org/wiki/Johnnetta_Cole. Accessed 19 Apr 2020
59. V. Lacapra, Interview: Mary-Dell Chilton on her pioneering work on GMO crops, Genetic Literary Project, 27 May 2015, St. Louis Public Radio. http://www.geneticliteracyproject.org/2015/05/27/interview-mary-dell-chilton-on-her-pioneering-work-on-gmo-crops/. Accessed 6 June 2015
60. National Inventors Hall of Fame, Inductees: Mary-Dell Chilton. http://invent.org/inductees/chilton-mary-dell/. Accessed 6 June 2015
61. The World Food Prize, Syngenta Scientist Dr. Mary-Dell Chilton Named 2015 National Inventors Hall of fame inductee, 5 May 2015. http://www.worldfoodprize.org/index.cfm/24667/35489/syngenta_scientist_dr_marydell_chilton_named_2015_national_inventors_hall_of_fame_inductee. Accessed 6 June 2015
62. Ada E. Yonath Facts. https://www.nobelprize.org/prizes/chemistry/2009/yonath/biographical/. Accessed 20 Apr 2020
63. Christiane Nüsslein-Volhard Biographical. https://www.nobelprize.org/prizes/medicine/1995/nusslein-volhard/biographical/. Accessed 20 Apr 2020

64. C. Nüsslein-Volhard. https://en.wikipedia.org/wiki/Christiane_N%C3%BCsslein-Volhard. Accessed 20 Apr 2020

65. J.A.C. Joselyn, Ph.D. https://www.cogreatwomen.org/project/jo-ann-cram-joselyn-phd/. Accessed 20 Apr 2020

66. J.A. Joselyn. https://en.wikipedia.org/wiki/Jo_Ann_Joselyn. Accessed 20 Apr 2020

67. S.B. McGrayne, *Nobel Prize Women in Science: Their Lives, Struggles, and Momentous Discoveries* (Carol Publishing Group, New York, 1993)

68. J.B. Burnell. https://starchild.gsfc.nasa.gov/docs/StarChild/whos_who_level2/bell.html. Accessed 20 Apr 2020

69. Science: Women of Impact, by Nadia Drake, Meet the Woman Who Found the Most Useful Stars in the Universe, 6 Sept 2018. https://www.nationalgeographic.com/science/2018/09/news-jocelyn-bell-burnell-breakthrough-prize-pulsars-astronomy/#close. Accessed 20 Apr 2020

70. J.B. Burnell. https://en.wikipedia.org/wiki/Jo. Accessed 20 Apr 2020

71. First Lady Hillary Rodham Clinton to Speak at Inaugural Gala for Rensselaer's 18th President, The Honorable Dr. Shirley Ann Jackson, Press Release, 17 Sept 1999. www.rpi.edu/dept/NewsComm/New_president/presshillary.htm. Accessed 23 Nov 1999

72. A.M. Perusek, Saluting African Americans in the National Academy of Engineering Class of 2001, SWE: Magazine of the Society of Women Engineers, Feb/Mar 2002

73. C. Pert. https://en.wikipedia.org/wiki/Candace_Pert. Accessed 21 Apr 2020

74. C. Pert, Ph.D. http://candacepert.com/. Accessed 21 Apr 2020

75. S. Malcom, Ph.D. https://usnewsstemsolutions.com/speakers/shirley-malcom/. Accessed 21 Apr 2020

76. S.M. Malcom. https://en.wikipedia.org/wiki/Shirley_M._Malcom. Accessed 21 Apr 2020

77. L.B. Buck – Biographical, Nobel Prize. https://www.nobelprize.org/prizes/medicine/2004/buck/biographical/. Accessed 21 Apr 2020

78. T. Grandin. https://en.wikipedia.org/wiki/Temple_Grandin. Accessed 21 Apr 2020

79. T. Grandin, Ph.D. https://www.templegrandin.com/. Accessed 21 Apr 2020

80. F. Barré-Sinoussi. https://www.nobelprize.org/prizes/medicine/2008/barre-sinoussi/facts/. Accessed 21 Apr 2020

81. F. Barré-Sinoussi. https://en.wikipedia.org/wiki/Fran%C3%A7oise_Barr%C3%A9-Sinoussi. Accessed 21 Apr 2020

82. F. Wong-Staal, Ph.D. https://history.nih.gov/nihinownwords/docs/transcripts/wongstaal.html. Accessed 21 Apr 2020

83. Biography of Flossie Wong-Staal. https://www.thewonderwomenproject.org/pages/biography-of-flossie-wong-staal. Accessed 21 Apr 2020

84. Biography of Lydia Villa-Komaroff. https://www.thewonderwomenproject.org/pages/biography-of-lydia-villa-komaroff. Accessed 21 Apr 2020

85. CEOSE – Member Biography. https://www.nsf.gov/od/oia/activities/ceose/Biographies/villa-komaroff.jsp. Accessed 21 Apr 2020

86. Women Scientist Profiles – Lydia Villa-Komaroff, Ph.D. https://womeninscience.nih.gov/women_scientists/villa-komaroff.asp. Accessed 21 Apr 2020

87. E.H. Blackburn, Facts, Nobel Prize. https://www.nobelprize.org/prizes/medicine/2009/blackburn/facts/. Accessed 21 Apr 2020

88. E. Blackburn. https://en.wikipedia.org/wiki/Elizabeth_Blackburn. Accessed 21 Apr 2020

89. About Anita Borg. https://anitab.org/about-us/about-anita-borg/. Accessed 21 Apr 2020

90. K. Mieszkowski, "Sisterhood is Digital," Fast Company, Sept 1999

91. President Clinton Names Anita Borg to the Commission on the Advancement of Women and Minorities in Science, Engineering, and Technology, White House Press Release, 29 June 1999. https://www.govinfo.gov/content/pkg/WCPD-1999-07-05/pdf/WCPD-1999-07-05.pdf

92. T. O'Brien, Women on the verge of a high-tech breakthrough, *San Jose Mercury News*, 9 May 1999

93. S. Eng, Women's group honors pioneers in technology, *San Jose Mercury News*, June 26, 1998

94. A. Borg, Women in Technology International Hall of Fame. https://www.witi.com/halloff-ame/102852/Dr.-Anita-Borg-Member-of-Research-Staff,-Xerox-PARC,-Founding-Director-Institute-for-Women-and-Technology/. Accessed 17 Aug 1999

95. Top 25 Women on the Web – Dr. Anita Borg. wysiwyg://5http://www.top25.org/ab.shtml. Accessed 17 Aug 1999

96. C.T. Corcoran, Anita Borg wants more scientists to start listening to women, Red Herring, Mar 1999

97. Method for quickly acquiring and using very long traces of mixed system and user memory references, Patent 5,274,811 granted 12/28/93. Patent 4,590, "Backup fault tolerant computer system, granted 20 May 1986

98. K. Hafner, Anita Borg, 54, creator of Systers list, *Rocky Mountain News*, 11 Apr 2003

99. Biography of Maria Klawe. https://www.hmc.edu/about-hmc/president-klawe/biography-of-president-maria-klawe/. Accessed 21 Apr 2020

100. M. Klawe. https://en.wikipedia.org/wiki/Maria_Klawe. Accessed 21 Apr 2020

101. About Harvey Mudd College. https://www.hmc.edu/about-hmc/. Accessed 21 Apr 2020

102. I. Daubechies. https://en.wikipedia.org/wiki/Ingrid_Daubechies. Accessed 21 Apr 2020

103. Duke Mathematician Awarded more than $400,000 for Her Contributions to Wavelet Theory. https://today.duke.edu/2018/07/duke-mathematician-awarded-more-400000-her-contribu-tions-wavelet-theory. Accessed 21 Apr 2020

104. Nomination of Susan Solomon https://www.nsf.gov/od/nms/recip_details.jsp?recip_id=332. Accessed 24 Apr 2002

105. Meet Susan Solomon. www.chemheritage.org/EducationalServices/faces/env/readings/Solomon.htm. Accessed 24 Apr 2002

106. S. Solomon: 1992 Common Wealth Award for Science and Invention. www.sigmaxi.org/prizes&awards/ssolomon.htm. Accessed 24 Apr 2002

107. S. Solomon. http://spot.colorado.edu/~gamow/george/1994bio.html. Accessed 24 Apr 2002

108. Susan Solomon explains Scott's fatal expedition to Antarctica, University of Leeds Press Release, 26 Sept 2001. www.leeds.ac.uk/media/current/Solomon.htm. Accessed 24 Aug 2002

109. NOAA Scientist Receives Nation's Highest Scientific Honor, NOAA News Online. www.noaanews.noaa.gov/stories/s368.htm. Accessed 24 Apr 2002

110. S. Solomon. https://eapsweb.mit.edu/people/solos/bio. Accessed 23 Apr 2020

111. F.H. Arnold. https://www.nobelprize.org/womenwhochangedscience/stories/frances-arnold. Accessed 21 Apr 2020

112. F. Arnold. https://en.wikipedia.org/wiki/Frances_Arnold. Accessed 21 Apr 2020

113. D. Strickland: Facts. https://www.nobelprize.org/prizes/physics/2018/strickland/facts/. Accessed 21 Apr 2020

114. D. Strickland. https://en.wikipedia.org/wiki/Donna_Strickland. Accessed 21 Apr 2020

115. C. Greider. https://www.nobelprize.org/womenwhochangedscience/stories/carol-greider. Accessed 21 Apr 2020

116. C.W. Greider. https://en.wikipedia.org/wiki/Carol_W._Greider. Accessed 21 Apr 2020

117. M-B. Moser, Facts, Nobel Prize. https://www.nobelprize.org/prizes/medicine/2014/may-britt-moser/facts/. Accessed 21 Apr 2020

118. M-B. Moser. https://en.wikipedia.org/wiki/May-Britt_Moser. Accessed 21 Apr 2020

119. M-B. Moser. https://www.ntnu.edu/employees/may-britt.moser. Accessed 22 Apr 2020

Appendix A: Scientific Occupations and Categories in this Volume [1]

Mathematics and Physical Sciences

Mathematics

Mathematician
Operations Research Analyst
Statistician, Mathematical
Mathematical Technician
Actuary
Statistician, Applied
Weight Analyst

Physical Sciences

Astronomy

Astronomer

Chemistry

Chemist
Chemist, Food
Toxicologist
Laboratory Supervisor
Chemical Laboratory Chief
Colorist
Perfumer
Chemical Laboratory Technician
Malt-Specifications-Control Assistant

Chemist, Instrumentation
Chemist, Wastewater-Treatment Plant
Assayer
Chemist, Water Purification
Laboratory Tester
Yeast-Culture Developer

Physics

Electro-optical Engineer
Physicist
Physicist, Theoretical

Geology*

Crystallographer
Geodesist
Geologist
Geologist, Petroleum
Geophysical Prospector
Geophysicist
Hydrologist
Mineralogist
Paleontologist
Petrologist
Seismologist
Stratigrapher
Engineer, Soils
Geophysical-Laboratory Chief
Geological Aide
Prospector
Paleontological Helper
Laboratory Assistant

Meteorology*

Meteorologist
Hydrographer
Oceanographer
Weather Observer

Other

Geographer
Geographer, Physical
Environmental Analyst
Materials Scientist
Aerial-Photograph Interpreter
Project Manager, Environmental Research

Laboratory Tester
Pollution-Control Technician
Test-Engine Operator
Tester
Criminalist
Photo-Optics Technician
Bottle-House Quality-Control Technician
Food Tester
Laboratory Assistant (utilities)
Cloth Tester
Laboratory Assistant (textile)
Pilot, Submersible

Computer Sciences

Software Engineer
Computer Systems Hardware Analyst
Quality Assurance Analyst
Computer Security Analyst
Data Base Administrator
Data Base Design Analyst
Microcomputer Support Specialist

Life Sciences

Agricultural Sciences

Agronomist
Animal Scientist
Dairy Scientist
Diary Technologist
Fiber Technologist
Forest Ecologist
Horticulturist
Poultry Scientist
Range Manager
Silviculturist
Soil Conservationist
Soil Scientist
Wood Technologist
Forester

Soil-Conservation Technician
Laboratory Technical, Artificial Breeding
Seed Analyst

Biological Sciences

Anatomist
Animal Breeder
Apiculturist
Aquatic Biologist
Biochemist
Biologist
Biophysicist
Botanist   .
Cytologist
Entomologist
Geneticist
Histopathologist
Microbiologist
Mycologist
Nematologist
Parasitologist
Pharmacologist
Physiologist
Plant Breeder
Plant Pathologist
Zoologist
Staff Toxicologist
Medical Coordinator, Pesticide Use
Food Technologist
Environmental Epidemiologist
Public-Health Microbiologist
Biology Specimen Technician
Herbarium Worker

Psychology

Psychologist, Developmental
Psychologist, Engineering
Psychologist, Experimental
Psychologist, Educational
Psychologist, Social
Psychometrist

Counselor
Counselor, Nurses' Association
Director of Counseling
Clinical Pshycologist
Psychologist, Counseling
Psychologist, Industrial-Organizational
Psychologist, School
Residence Counselor
Vocational Rehabilitation Counselor
Psychologist, Chief
Clinical Therapist
Counselor, Marriage and Family
Substance Abuse Counselor
Director of Guidance in Public Schools

Other

Park Naturalist
Feed-Research Aide
Vector Control Assistant
Biological Aide

Social Sciences

Economics

Economist
Market-Research Analyst
Director, Employment Research and Planning

Political Science

Political Scientist

History

Biographer
Director, State Historical Society
Genealogist

Historian
Historian, Dramatic Arts
Director, Research

Sociology

Research Worker, Social Welfare
Sociologist
Clinical Sociologist

Anthropology

Anthropologist
Anthropologist, Physical
Archeologist
Ethnologist
Conservation, Artifacts

Other

Philologist
Scientific Linguist
Intelligence Research Specialist
Intelligence Specialist

Other scienc-related occupations are found in the categories "Architecture, Engineering, and Surveying" and "Medicine and Health."

* Geology, geophysics, meteorology, and oceanography are sometimes referred to as the Earth Sciences [2].

References

1. U.S. Department of Labor, *Dictionary of Occupational Titles*, 4th edn. Revised 1991, Volume 1
2. C.A. Ronan, *Science: Its History and Development among the World's Cultures* (The Hamlyn Publishing Group Limited, New York, 1982)

Index

Printed in the United States
by Baker & Taylor Publisher Services